Education & Professional Development of Engineers in the Maritime Industry 2016

Singapore
20-21 September 2016

ISBN: 978-1-5108-8334-5

Printed from e-media with permission by:

Curran Associates, Inc.
57 Morehouse Lane
Red Hook, NY 12571

Some format issues inherent in the e-media version may also appear in this print version.

Printed with permission by Curran Associates, Inc. (2026)

For permission requests, please contact The Royal Institution of Naval Architects at the address below.

The Royal Institution of Naval Architects
8-9 Northumberland Street
London, WC2N 5DA
United Kingdom

Phone: 020 7235 4622
Fax: 020 7259 5912

publications@rina.org.uk

Additional copies of this publication are available from:

Curran Associates, Inc.
57 Morehouse Lane
Red Hook, NY 12571 USA
Phone: 845-758-0400
Fax: 845-758-2633
Email: curran@proceedings.com
Web: www.proceedings.com

CONTENTS

INTERNATIONAL COLLABORATIVE WORK TO IMPROVE RESEARCH QUALITY AND ENHANCE ACADEMIC ACHIEVEMENT

I K A P Utama, Institute of Technology Sepuluh Nopember (ITS), Surabaya, Indonesia
B Ganapathisubramani, University of Southampton, Southampton, UK
N Hutchins, University of Melbourne, Melbourne, Australia
B Nugroho, University of Southampton, Southampton, UK
J P Monty, University of Melbourne, Melbourne, Australia
F A Prasetyo, PT Biro Klasifikasi Indonesia (Persero), Jakarta, Indonesia
M Yusuf, PT Dharma Lautan Utama (DLU), Surabaya, Indonesia
M Tullberg, Hempel A/S, Copenhagen, Denmark

SUMMARY

A large scale joint research project is underway, with the aim of improving the efficiency of ship operations by providing a methodology and technology that can accurately quantify the emission and fuel usage penalty due to bio-fouling of the hull. The research involves six different institutions from four countries (Australia, Denmark, Indonesia, and UK) bringing together universities, a passenger ship company, a manufacturer of anti-fouling coatings, and the Indonesian Classification Society. Here the importance of individual role and contributions from each institution are discussed. The six participants in general represent academic, industrial, and an independent party that promote of safety of life, property and the environment as well as to support policy makers, each with different objectives and interests that are interconnected. This paper will explains the importance of collaboration and satisfying individual objectives from each field in order to achieve the collaborative overarching aim

1. INTRODUCTION

In the last decade the issue of global warming and the energy crisis have prompted efforts to find methods of reducing energy consumption for shipping industries (Davies, 2006; Longva et al., 2010). Countries such as Australia, Indonesia, and the United Kingdom (UK) all rely heavily on long-haul shipping transportation due to their unique geographical location and condition. Australia is a large continent, with diverse population centres, that is geographically isolated from major trading partners. In addition, the strong resource sector puts strong demands on export bulk shipping. Indonesia is an archipelago of 17,499 islands with a large population (250,000,000). The country is heavily reliant on ships and ferries to transport people and goods between islands. With an average of 4-5% economic growth (Tabor, 2015), Indonesia's dependence on naval transport will rise significantly. UK is geographically separated from the main European continent by the English Channel and the North Sea. The country is also heavily reliant on naval transport to import and export goods. Faced with the ever increasing pressure to reduce the environmental, energy, and economic footprint, it is important for countries such as Australia, Indonesia, and the UK to address energy issues in the shipping industry.

Hull roughness is an important, but largely unquantifiable contributor to the overall energy expenditure and emissions from the worlds shipping fleet. The hull of a ship has underlying roughness due to manufacturing imperfections (weld seems, panel misalignment etc), coatings (topology resulting from application of corrosion and foul-resist coatings) and most importantly from Biofouling: the settlement of marine organisms on to the ship hull. Many studies, particularly by Schultz (2004, 2007, 2011) have shown that biofoulings have significant contribution to the increase of ship's drag, with a large economic penalty. Despite the severity of this issue, currently there are no reliable methods to enable ship operators to quantify the drag penalty due to hull roughness for an in-service vessel.

Considering the global scale of the challenge, an international research collaboration that involves six institutions from four different countries has been formed, namely: Institute of Technology Sepuluh Nopember, Biro Klasifikasi Indonesia (Indonesian classification society), and Dharma Lautan Utama passenger ship Company from Indonesia; the University of Melbourne from Australia, , the University of Southampton from the United Kingdom (UK); and Hempel A/S from Denmark.

The main objective and overarching aim of the collaborative research is to understand the skin friction drag on ships hulls due to biofoulings and to improve the efficiency of ship operations by providing a methodology and technology that can accurately quantify the emission and fuel usage penalty. This can be obtained through detailed under-water image-based surface scanning techniques, coupled with in-situ laser-based measurements of the water flow over the hull of the ship. These methods will provide direct data on the increase in drag experienced by the ship under different fouling conditions. The two shipboard experiments are complemented and validated by detailed laboratory experiments. The outcome of the investigation will permit the maritime industry and policy makers to make more informed decisions in terms of operations at operational or regulatory level. The research collaboration also allows the academic institutions to improve research quality and enhance academic achievement and student outcomes.

To accomplish the main objective, the six different institutions need to work and collaborate closely, particularly because each of them has different expertise and specific roles that complement each other. In general the six participants can be divided into *three major fields*, namely: academic, industry, and classification society to support policy maker. Each of these fields has different interests and expected outcomes, however they form the base of the overall main objective for the research collaboration. In this report we will discuss the motives and details of the joint research, the role of each institution, the expected outcome from individual fields, and how these individual objectives are correlated and satisfied to achieve the overall goal

2. MOTIVE OF FORMING RESEARCH COLLABORATION

The proposed collaborative research project requires a large quantity of facilities, funding, and personnel due to its scale. To the best of author's knowledge there have been very few attempts at this type of investigation. However, one such study is by Lewthwaite et al. (1984). Their research involved three different institutions, namely: a university, a government research agency, and a marine company. Hence, this type of research collaboration is a very challenging venture to be performed individually. Therefore it is desirable to share the research burden with various institutions that have similar aspiration (Katz and Martin, 1997). Studies regarding research collaboration show that scientific endeavour is becoming more collaborative, particularly between different organisations, both within a country or internationally (Subramanyam, 1982; Melin and Persson, 1996). The same studies also show that collaboration improves scientists' visibility and productivity.

Although collaboration between academic institutions is very common, the amount of collaboration between academic institutions and industry or government bodies is comparably lower. According to Bruneel et al (2010) there are several barriers that impact research collaboration between the different fields of academia, industry, and policy makers. Particularly in the form of differing methods of administering private and public knowledge (Bruneel et al, 2010; Dasgupta and David, 1994). At its core, scientists in academic institutions are motivated by academic excellence and education while industry firms are driven by economic value derived from competitive advantage (Dasgupta and David, 1994).

Despite these challenges, Bruneel et al (2010) reveal that inter-organisational trust is needed to lower barriers between academic institutions and industry. They even go further by implying that informal exchange and interaction should be attempted to build good rapport. We believe that a similar method can be applied when collaborating with government institutions. For this research collaboration, it is clear that close cooperation and communication is needed. Particularly when the possible benefit for all parties involved clearly outweighs any drawbacks.

3. THE COLLABORATIVE RESEARCH METHODOLOGY

The research methodology can be subdivided into four research strands (or work packages), which will meet the overall aim of the project.

3.1 RESEARCH STRAND 1: LASER DOPPLER ANEMOMETER MEASUREMENT (LDA).

For this research strand we intend to perform direct in-situ measurements using LDA in the boundary layer formed over a ship's hull under sailing conditions, to monitor the influence of the accumulation of bio-fouling on the skin friction drag (see Buchhave et al 1979 and Tropea 1995 for further reading regarding LDA). Such direct investigations have been very rarely performed, but are critical if we are to be able to ascribe a precise figure to the economic impact of bio-fouling to ship operations. One notable attempt at a similar study is that by Lewthwaite et al. (1984), however they used pitot tubes instead of LDA, and did not obtain a record of the hull state.

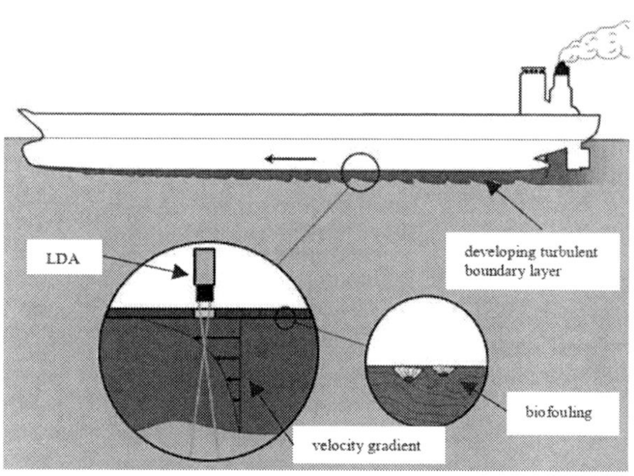

Figure 1 : Illustration of the LDA with turbulent boundary layer and surface roughness in ships

Here a small window is placed in the hull of an Indonesian operating ship during dry-docking, at which point the ship is also cleaned and recoated with anti-fouling products. The LDA is installed in the hull of the ship to measure the velocity gradient in the turbulent boundary layer formed over the ship's hull as it sails (see figure 1). The LDA can obtain an in-situ fluctuating velocity signal across some traversable distance from close to the hull surface, to at least the end of the logarithmic region. From the velocity gradient we can directly calculate the percentage increase in skin friction drag. This will be monitored (along with other key data - GPS coordinates, velocity, sea-state, fuel usage, draft etc) during the course of the inter dry-docking period (1 - 3 years depending on funding level).

3.2 RESEARCH STRAND 2: IMAGE BASED SURFACE SCANNER

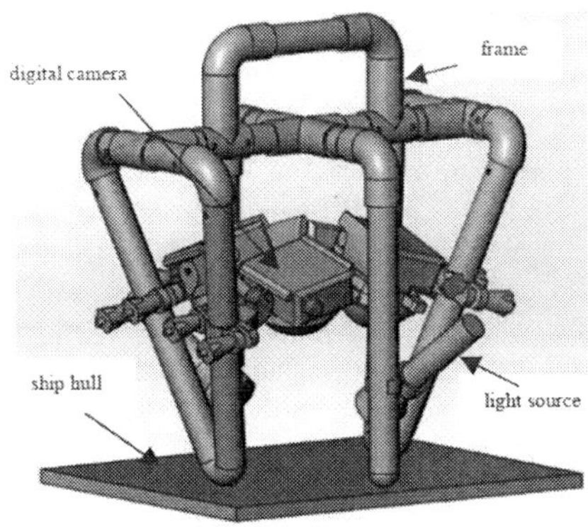

Figure 2 : Surface scanner

One thing that lacking from Lewthwaite et al.'s (1984) report is the absence of the hull-condition record. The unavailability of this data prevented them from quantifying the effect of biofouling topography to the skin friction drag in more detail. To avoid similar drawbacks here we use an image-based surface scanner to monitor the state of the hull surface finish. This activity starts during dry-docking, when the quality of the surface finish from the spraying and cleaning process will be quantified and will continue through regular (approx monthly) dive inspections until the next dry-docking.

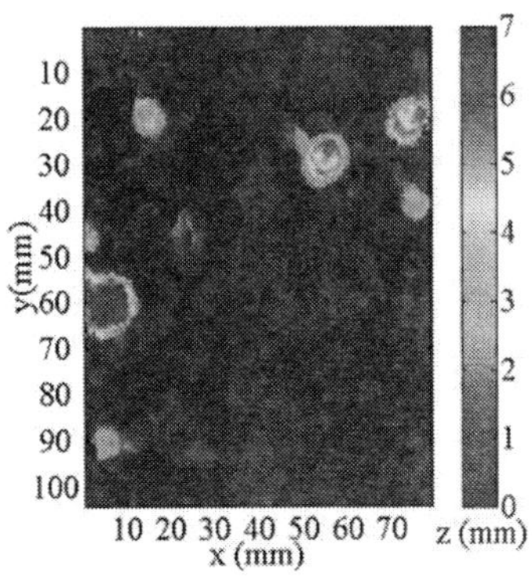

Figure 3: Digitally reconstructed biofouling from surface scanner

Figure 2 shows the proposed surface scanner. It is made of four digital underwater cameras attached frame. The previously calibrated four cameras will simultaneously take a photo of the hull surface and then digitally reconstruct (see figure 3) the surface topography. The scanning of the bio-fouled hull is performed alongside the measurement with the LDA system, which allows a direct comparison between the physical dimension of the bio-fouling and its effect on the ship skin-friction drag. The scanned data will later be used in the third and fourth strands of this research.

3.3 RESEARCH STRAND 3: DATA CORRELATION

Here the LDA measured drag, and ship-board data such as GPS coordinates, velocity, sea-state, fuel usage, draft etc are mined and correlated with the surface scan data. This will allow us to produce a full scale prediction of drag penalty and in an attempt to produce a workable correlation/equation that will ultimately enable the hull fouling penalty to be calculated directly from surface scan information.

3.4 RESEARCH STRAND 4: AERODYNAMICS TESTING

For this strand we intend to investigate the effect of a realistic bio-fouling on the turbulent boundary layer under more controlled laboratory conditions to validate the efficacy of the proposed LDA gradient-based technique. The LDA method is novel and unproven. There have been many studies that investigate the effect of bio-fouling on turbulent boundary layers. Some of these studies involve an actual bio-fouling that has been grown in their natural habitat and then removed to a laboratory for experiment (Walker et al (2013), Schultz and Swain (1999)). However, this technique is relatively difficult to perform due to the challenges in preserving them for lab experiments, particularly the effect of change in the natural environment (from ocean to lab water tank). One method to avoid this challenge is by realistically replicating the bio-fouling through casting, machining, rapid-prototyping or combinations of the three. The advantages of replicated geometries are that it is not necessary to introduce fouling into sensitive experimental facilities and the surfaces can be scaled for testing at different conditions or in the case of bio-fouled surfaces, testing using fluids other than seawater (i.e fresh water or air).

Using the digitally reconstructed and scaled biofouling data from the surface scanner (from strand 2), a detailed master model is manufactured and replicated using a CNC-Machine. To produce multiple replicas with sufficient quantities to fill the working section of a wind tunnel, copies of the master tile are made via moulding and casting techniques. A similar method was used by Nugroho et al (2013, 2014) for their surface roughness experiment.

The end result of this planned project is a catalogue of data taken over a certain period of time to observe the build-up of the bio-fouling under normal operations and also to measure the changes in the skin-friction drag.

4. RESEARCH GROUP FORMATION AND ROLE

The large scale research project has a clear main objective, which is to understand the skin friction drag on ships hulls due to biofouling and to improve the efficiency of ship operations by providing a methodology and technology that can accurately quantify the emission and fuel usage penalty. In order for this endeavour to succeed, it requires the involvement of many different institutions, from academia, industry, to government representatives. Each of these institutions has their own specific role and expertise in order to meet the overarching aim. Here we will discuss the research group formation process and role for each of the stake-holders.

The research project is initially proposed by the University of Melbourne and Institute of Technology Sepuluh Nopember (ITS) through the newly formed Australian-Indonesian Centre (AIC) program in late 2014. The aim of the program is to establish and strengthen collaboration between Australian and Indonesian universities. As one of Indonesia's leading universities, ITS has a very strong marine technology faculty and strongly linked to various marine intuitions. The main role of ITS is to host the field experiment, use their extensive contacts in the maritime industry, assist with the import of various scientific apparatus from overseas, and liaise with local government representatives and recruit Masters students onto this project.

In order to allow the project to move forward ITS has invited PT Dharma Lautan Utama (DLU), an Indonesian passenger ship company, to be involved in the research project. The ship company has kindly provided the research team with access to a RO/RO passenger ferry called Dharma Kencana IX, as the test bed. The ferry is in operational status and serves between Merak and Bakaehuni, one of Indonesia's busiest lines. To ensure the safety of the proposed ship experiment for the entire duration of the investigation, ITS also invited Biro Klasifikasi Indonesia (BKI), Indonesian classification society, to be part of the research team. BKI is a recognized organization by Indonesian Government that is tasked by the Government to classify any Indonesian flagged marine transport both domestic and international that operates in Indonesian water and/or world wide voyage [DJPL, 2016]. They provide contribution in the form of technical evaluation regarding ship design, modification, reconstruction, maintenance and other dry-docking activities, in accordance with BKI technical standards (rules) for design and construction of ships and Indonesian Government maritime regulations.

As the initiator of the research project and part of the AIC research program, the University of Melbourne has several important roles. The first is to manage the overall research schedule. Secondly, the University's major contribution is in the form of field research equipment, such as the LDA, surface scanner, and other supporting research apparatus. Finally, they also host the laboratory investigation due to the institution's extensive experience in aerodynamic wind tunnel experiments. The University of Melbourne has a very strong fluid mechanics background and extensive laboratory facilities, furthermore it is relatively close to Indonesia. Therefore, it is an ideal location for the laboratory research.

Considering the scale of the experiment, the University of Melbourne invited Hempel A/S, a multinational antifouling company that is well versed in the research of biofouling to be part of the investigation. The company contributes in the form of antifouling technology that is applied on the hull of Dharma Kencana IX during dry docking when the hull is in clean state. Our main field of research activity is to measure the change of fluid flow on a ship's hull due to biofouling growth. However, the LDA method used on the ship is novel and unproven, hence it requires many repeated tests under constant conditions. Therefore, it is desirable that the ship hull's condition is unchanged during this time. The anti-fouling paint from Hempel A/S prevents biofouling from attaching and growing on the ship's hull for a longer period of time, permitting the field researchers to test the LDA during the key initial stages of this project. Furthermore, Hempel A/S also offers their technical expertise, industry contacts and extensive experience of the maritime industry.

Although at this stage the research team had secured the necessary facilities and equipment for the project to move forward, they still lacked the necessary personnel to perform the field and laboratory experiment. To overcome this challenge, ITS and the University of Melbourne invited The University of Southampton as part of the research team. The University of Southampton has a long history in naval and fluid mechanics research. The University of Melbourne and the University of Southampton have previously performed similar laboratory experiments, where they scanned a sample of tubeworm type fouling and characterised the drag using a wind tunnel (see Monty et al, 2016). Therefore the University of Southampton is well versed in this type of investigation.

In early 2015 ITS and the University of Southampton managed to secure funding from the UK government through the Newton fund program administered by the British Council, leveraging the previous funding from the AIC. The new funding allows the University of Southampton to recruit the necessary personnel to perform the field and laboratory experiment. Furthermore it allows the UK and Indonesian universities to fund the field travel expenses.

5. CORRELATION BETWEEN INDIVIDUAL FIELD OBJECTIVES

The challenge in this research collaboration is to satisfy each field's individual objective without sacrificing the overall goal. Here we will discuss each field's objective and how they connect with each other.

5.1 ACADEMIC FIELD OBJECTIVE

In this research collaboration, the three academic institutions: ITS, the University of Melbourne, and the University of Southampton, have two targets, namely: to obtain further understanding about turbulent boundary layer flows over rough surfaces and to educate and train higher degree students.

First and foremost, although it is well known that biofouled hulls cause an increase of skin friction drag, leading to increases in power and fuel usage, our understanding of the turbulent flow over rough walls (such as a fouled hull) is still lacking. One of the main challenges in studying wall bounded turbulent flow, even over a smooth wall, is the seemingly quasi-random and complex motions inside the turbulent boundary layer. This challenge is exacerbated when the flow moves over uneven surfaces such as a fouled hull. The combined in-situ and laboratory experiments proposed will allow further understanding about the rough-walled turbulent boundary layer, particularly for realistic roughness. To the best of the author's knowledge this type investigation has never been performed before. The expected outcome from the project is a database with high research value. These results will be published in peer-reviewed journals and disseminated at conferences, allowing the three academic institutions to communicate their results to the scientific communities.

Apart from the research endeavour, this project provides opportunities for the academic institutions to train and educate graduate students, and also to provide crucial international / global experiences. The University of Melbourne uses this opportunity to design a capstone project that requires four Masters Students to perform an initial study for the image based surface scanner. The activity allows the students to master digital image reconstruction techniques and to apply their technical skill and engineering knowledge. The capstone Masters project also serves as an initial study for the Postdocs from the University of Southampton to continue and expand.

The research collaboration is particularly beneficial for ITS, as it allows them to train two Masters students. The two students come from a program called "Pre to Post MSc", funded by the Ministry of Research, Technology and Higher Education. The aim of this program is to train and educate high achiever graduate students at ITS from other smaller or younger Universities that are located in the eastern part of Indonesia. The students are later expected to return to their respective universities and work as university lecturers. The research project allows ITS to fund the student's travel expenses to the field research site and have them experience high level international research collaboration. This project also allows ITS to prepare more high achiever Masters students to be involved, particularly those aiming to continue their study in PhD at The University of Melbourne and The University of Southampton. These graduates are also expected to return to ITS as lecturers and contribute as part of the effort by ITS to be a world-class research university.

For ITS, the collaborative project also represents the three core roles for higher education in Indonesia, namely (1) education and teaching, (2) research and development of science, technology and arts, (3) community services. The three philosophies are more commonly known as TRI DHARMA PERGURUAN TINGGI, which should be developed simultaneously (Utama and Pribadi, 2011).

5.2 INDUSTRIAL FIELD OBJECTIVE

The collaborative research project involves two industrial representatives: PT DLU (passenger ship company) and Hempel A/S (manufacturer of antifouling coatings). The two firms are from different industries; however they are related and complement each other. As a passenger ship company, PT DLU earns revenues from transporting passengers (including trucks and cars) between islands in Indonesia. To boost profit, many ship companies require an effective and efficient ship operation management. Two of their largest sources of expenses are fuel and dry docking. Fouled hulls of ships results in higher fuel usage than a clean one, which translates to greater expenditure and emissions. In order to remove the biofouling from the hull and return it to the smooth state, the ship must undergo a dry docking process, which has a considerable financial cost. Furthermore, during dry docking the ship is unable to generate revenues, resulting in further losses. Note that the interval of dry docking varies for each type of ship on the basis of class society rules and administrator regulation. Passenger ships have to perform dry docking once a year (DJPL, 2014; BKI, 2016). During dry docking the ship's hull is cleaned, the underwater hull of the opening and closures in the shell related to the machinery system is checked and inspected. The ship hull is also protected with anti-fouling paint, a product that companies such as Hempel A/S produce.

For PT DLU, having a more informed decision in terms of operations will be beneficial. The proposed LDA device could eventually form part of the standard sensor suite on a modern ship, providing a constantly updating report on the health of the hull and the economic ramifications of the deteriorating hull smoothness. Furthermore, the surface scanner technique also has potential to be commercialised. In the future, ship operators and anti-fouling companies can use the surface scanner and database catalogue of various biofouling to predict a ship's excess fuel usage. The project also serves as a good opportunity for Hempel A/S to test their product and analyse the anti-fouling performance from a fluid dynamics perspective.

5.3 POLICYMAKER FIELD OBJECTIVE

As an institution that is tasked to classify any Indonesian flagged marine transport that operates in worldwide water and specially, Indonesian waters, BKI has an important role in advising the Indonesian Government regarding various marine safety, environmental protection, and procedures. One of the major outcomes of this research project is more detailed information regarding the energy and economic ramifications of a fouled ship hull. The information from the research can be used by BKI as a basis to review Government policy regarding the management of energy efficiency and biofouling treatment procedure. This would allow more economical and efficient ship operations. Furthermore, the policy can also be applied internationally through the International Maritime Organisation (IMO), where Indonesia is an active member. For developing countries such as Indonesia, research based policy is difficult to perform due to the country's low national research budget. Therefore this international research collaboration is a good opportunity to apply and integrate the research results into policy.

6. CONCLUSIONS

A large scale research collaboration between various institutions from four different countries that involves representatives from academia, industry and government policy makers is currently underway. The overall aim of the project is to understand the negative effect of growth of biofouling on a ship's hull in more detail and to come up with a method and technology that can quantify the energy usage penalty in order to improve the efficiency of ship operations.

Each of the six institutions involved have different roles and expertise that complement each other. The three academic institutions provide scientific expertise and apparatus. The two industry firms, a passenger ship company and manufacturer of anti-fouling solutions, have supplied the ship as a test bed and the anti-fouling technology respectively. Finally, the classification society provides technical advice, ensures the safety of the project, and provides avenues for the research to impact government policy. By collaborating together they leverage the value of the research viand are able to access funding from the Australian, Indonesian, and UK government.

Here we also have shown that despite the research team coming from various different fields, it is possible to form a joint project together. Each field (academia, industry, and policy maker), has different interests and individual objectives that are specific to their needs. The three academic institutions are interested to gain further understanding in rough wall turbulent boundary layers, to educate higher degree students, and to see their research impact a broader society. The two industrial firms are interested with the possible outcome of more efficient operations and new technology that can be applied. Finally, the Classification Bureau can provide a better policy based on the research outcomes. From the individual objectives of each field, one can immediately see that they are complementary and synergistic, making this a well-structured project By satisfying the requirements of each field, we are confident that the overarching aims of this project can be obtained.

7. ACKNOWLEDGEMENTS

The authors would like to thank the Australian Research Council (ARC), the Australia Indonesia Center (AIC) and the Newton Fund from the British Council for financial support.

8. REFERENCES

1. BIRO KLASIFIKASI INDONESIA, BKI, (2016), Rules for Classification & Surveys, Volume 1, Part 1.

2. BRUNEEL, J., D'ESTE, P., SALTER, A., (2010), Investigating the factors that diminish the barriers to university–industry collaboration. *Research Polic.* 39: 858 -868

3. BUCHHAVE, P., GEORGE, W. K. JR., LUMLEY, J. L., (1979). The measurement of turbulence with the laser-doppler anemometer. Ann. Rev. Fluid Mech 11:443-503

4. DASGUPTA, P., DAVID, P., (1994). Towards a new economics of science. Research Policy, 23: 487–522.

5. DAVIES, M., (2006) Emissions trading for ships- a european perspective. Naval Eng. J. 118:3, 131– 138.

6. Direktorat Jenderal Perhubungan Laut, DJPL (2014), Peraturan Direktur Jenderal Perhubungan Laut Nomor. HK.103/I/4/DJPL-14 tentang Pengedokan (Pelimbungan) Kapal Berbendera Indonesia.

7. Direktorat Jenderal Perhubungan Laut, DJPL (2016), Surat Keputusan Direktur Jenderal Perhubungan Laut Nomor PK.204/1/3/DJPL-16 tentang Penunjukan Biro Klasifikasi Indonesia (Persero) sebagai Recognized Organization untuk melaksanakan survey dan sertifikasi Statutoria atas nama Pemerintah pada kapal Berbendera Indonesia.

8. KATZ,J. S., MARTIN, B. R., (1997) What is research collaboration?. *Research Policy*. 26: 1 - 18

9. LEWTHWAITE. J. C, MOLLAND. A. F, THOMAS. K. W. (1984) An investigation into the variation of ship skin frictional resistance with fouling. Trans RINA 127:269–284.

10. LONGVA, T., EIDE, M. S., SKJONG, R., (2010) Determining a required energy efficiency design index level for new ships based on a cost-effectiveness criterion. Maritime Policy and Management. 37:2, 129–143.

11. MONTY, J. P., DOGAN, E., HANSON, R., SCARDINO, A. J., GANAPATHISUBRAMANI, G., HUTCHINS, N., (2016) An assessment of the ship drag penalty arising from light calcareous tubeworm fouling. Biofouling, 32(4): 451-64

12. MELIN, G., PERSSON, O., (1996) studying Research Collaboration Using Co-authorships. *Scientometrics*. 36 (3): 363 - 377

13. NUGROHO, B., HUTCHINS, N., MONTY, J. P., (2013) Large-scale spanwise periodicity in a turbulent boundary layer induced by highly ordered and directional surface roughness. Int J Heat Fluid Flow, 41:90 -102.

14. NUGROHO, B., GNANAMANICKAM, E. P., KEVIN, MONTY, J. P., HUTCHINS, N., (2014) Roll-modes generated in boundary layers with passive surface modification. AIAA Paper 2014-0197

15. SUBRAMANYAM, k. (1982) Bibliometric studies of research collaboration: A review. *Journal of Information Science*. 33 - 38

16. SCHULTZ, M. P. (2004) Frictional resistance of antifouling coating systems. J. Fluids Eng. 126, 1039.

17. SCHULTZ, M. P. (2007) Effects of coating roughness and biofouling on ship resistance and powering. Biofouling 23(5-6), 331–341

18. SCHULTZ, M. P., BENDICK, J. A., HOLM, E. R. & HERTEL, W. M. (2011) Economic impact of biofouling on a naval surface ship. Biofouling 27, 87–98.

19. TABOR, S. R. (2015) Constraints to Indonesia's Economic Growth. ADB Papers on Indonesia, no 10.

20. TROPEA, C., (1995) Laser doppler anemometry: recent developments and future challenges. Meas. Sci. Technol. 6 : 605-619.

21. Utama, I K A P and Pribadi, T W (2012). *Three in One Education Concept, an Indonesian Perspective*. Education and Professional Development of Engineers in the Maritime Industry, 7th – 8th December 2011, Newcastle University, UK.

22. WALKER, J. M., Sargison, J. E., and Henderson, A. D., (2013) . Turbulent boundary-layer structure of flows over freshwater biofilms. Exp. Fluids, 54:1–17.

9. AUTHORS BIOGRAPHY

Prof I Ketut Aria Pria Utama is a professor of ship hydrodynamics at the Institut Teknologi Sepuluh Nopember (ITS) at Surabaya, Indonesia. He obtained his PhD in 1999 from the University of Southampton. He is a fellow member of the Royal Institution of Naval Architects (RINA) since 2006 and member of the Indonesian Academy of Sciences (AIPI) since 2015. His research interests include ship resistance and propulsion, seakeeping, computational fluid dynamics, renewable energy, and maritime engineering education.

Prof Bharathram Ganapathisubramani is a Professor of Experimental Fluid Mechanics in the Aerodynamics and Flight Mechanics Research Group at the University of Southampton. He was previously a Senior Lecturer at Southampton (2010-2012) and a Lecturer at Imperial College London (2007-2010). There are three major strands in BG's research and he has made several unique contributions in each of them. The first strand is aimed at physics and control of turbulent flows in aero-/hydro-dynamic applications. Examples include quantifying the effects of scale interactions in turbulent shear flows, isolating large-scale mechanisms responsible for skin-friction drag and scaling of large-scale energy containing motions in smooth-wall turbulent boundary layers, effects of geometric modifications of bluff bodies on dynamics of turbulence and established scaling laws and using multi-scale geometries for flow control. The second strand is focussed on fluid dynamics of biological/bio-inspired systems. Work in this area includes CT-scan based simulation of fluid flow through lymphatic node, aeromechanics of active and passive membrane wings, evolution of flight and swimming performance of marine reptiles. The final strand, which serves as the link between the other strands of my research, is on development of new experimental and data reduction methods including pressure determination using planar and volumetric velocimetry data, temperature measurements in high-speed flows and innovative ways of examining the data.

Dr Bagus Nugroho is a Research Fellow in Department of Engineering and Environment at The University of Southampton and visiting Research Fellow in Department of Mechanical Engineering at The University of Melbourne. Dr Nugroho's research interests are in wall-bounded turbulent flow, passive flow control, drag reduction mechanism, and flow over rough surface. He also looks into environmental fluid flow, particularly at the effects of biofoulings on marine engineering applications.

A/Prof Jason Monty is an ARC Future Fellow and Associate Professor in Mechanical Engineering at The University of Melbourne. A/Prof Monty is an experimental researcher in Fluid Mechanics and is the leader of the Michell Hydrodynamics Laboratory at Melbourne, which houses several new and unique flow facilities and state-of-the-art instrumentation. A/Prof Monty has a broad range of research interests from arterial flows to wall-turbulence to wave mechanics. His major contributions to science have been in the field of turbulent boundary layers, which has led to new experimental programs in turbulent air-sea interactions and in-situ measurements of the boundary layer on a ship.

A/Prof Nicholas Hutchins has been an ARC Future Fellow and is an Associate Professor in Mechanical Engineering at The University of Melbourne. A/Prof Hutchins is an experimental researcher in Fluid Mechanics, specialising in turbulent boundary layers, rough wall flows, flow control of turbulent boundary layers, drag reduction and advanced flow diagnostics. His major contributions have been in the classification of large-scale structures in turbulent boundary layers, and also in quantifying errors in fluid measurement techniques.

Dr Fredhi Agung Prasetyo holds the current position as Senior Manager of Research & Technical Application Division Research & Development, Biro Klasifikasi Indonesia. He has main research interest on fatigue & fracture, corrosion protection, analysis on ocean environmental effect to the maritime & ship safety and environmental protection.

Mr Mohammad Yusuff is the Director of Armada, Engineering, and Logistic for PT Dharma Lautan Utama. Mr Yusuf has an extensive 16 years of experience in engineering, operation, and maintenance for various types of passenger ship.

Dr Marcus Tullberg holds the position of Team Leader, Hull Performance in R&D at Hempel A/S. He has more than eight years of experience in developing antifouling coatings as well as analysing the hull and propeller performance of ships using on-board measured data.

MARITIME DEGREES AND PROFESSIONAL DEVELOPMENT THROUGH DISTANCE E-LEARNING: A CASE STUDY OF AN INTERNATIONAL LEARNING ALLIANCE

R Thain, Chief Executive and Academic Director, Marine Learning Alliance

J Cross, Science Education Specialist, Marine Learning Alliance

D Kelly, Director, AsiaPac The Institute of Marine Engineering, Science and Technology (IMarEST)

SUMMARY

The global shortage of engineers in all disciplines has been well documented, and existing provision of engineering programmes within the traditional full-time residential university sector is not keeping up with demand. For many prospective students, or working engineers who are required to undertake continuous professional development to maintain their professional registration, full-time classroom based attendance for education and training can be an issue. Financial, geographical or time constraints frequently create barriers to participation. The growth of niche, specialist providers of education using innovative methods of design and delivery can provide a viable and effective solution to these issues. The Marine Learning Alliance (MLA) brings together industry, academia and a professional body to develop and deliver Bachelors and Master's degrees, and also professional development courses which are highly relevant to the needs of maritime employers. Using a technology-enhanced delivery platform which provides 'anytime anywhere' access to learning materials, MLA has developed methods of design, production and delivery which offer the multimedia-rich experience that today's student seeks, whilst maintaining academic quality and standards. This is a critical balance to achieve, and requires much more effort than simply repurposing existing content from residential courses into a form of e-learning. This paper presents a case study of a highly successful international learning alliance. The experience of designing and delivering Master's degree and professional development courses by distance e-learning is discussed, together with the benefits which can be afforded by such a model.

1. INTRODUCTION

Recent studies indicate an emerging skills gap emerging in engineering. In 2015, 53% of businesses in the UK expected difficulties in recruiting STEM-skilled staff (Science, Technology, Engineering, Mathematics), and 48% of engineering enterprises said that hard-to-fill vacancies meant delays in next products and services [1]. The UK is not alone in the struggle for skilled engineers and technicians, with the recent Manpower Global Talent Shortage Survey [2] showing that the top three shortages globally are skilled trades, engineers, and then technicians (Figure 1). This is a problem which carries through into the marine and maritime world, with recent studies indicating a potential shortfall of around 3,500 trained UK deck and engineer officers at sea by 2021 [3]. These deck and engineering officers, in common with other seafaring professions such as hydrographic surveyors, conduct high-value, mission critical work which requires a significant level of education and training. This education and training aspect usually begins with a programme of further education (FE) or higher education (HE), traditionally undertaken on a residential attendance basis. Following employment and work experience, participating in on-the-job training, and undertaking post-qualification continuing professional development (CPD), an individual eventually receives professional body certification (often through a Chartered descriptor) as a 'licence to practice'.

Delivering more STEM personnel into the workforce through this well-established route may not be the only way to deal with the growing skills gap issue. As the employment landscape evolves, employees no longer expect a 'job for life', and are increasingly seeking to gain new skills and qualifications throughout their careers; Higher Education has thus become relevant *for* working life as well as contributing to the development *of* working life [4]. STEM employees may cross between disciplines, and in such cases, significant cross-training and re-skilling could be required. In addition, the STEM workplace is characterised by the necessity for frequent change in response to developments in technology and professional practice and, therefore, there is a demand from both employers and employees for regular upskilling throughout the employment pathway. HE and FE institutions, employers and professional bodies all have a role to play in ensuring that provision of education and training is adequate to meet demand, and that the workforce has timely opportunities to develop within their careers to suit personal and professional needs.

Distance learning represents an opportunity for students of all ages and experiences to work and learn at their own pace, almost anywhere in the world and it offers the new job market entrant and more experienced personnel huge opportunities to remain current and progress their careers through further study, whilst maintaining personally satisfying 'work life' balances, at a reasonable cost. Technological enhancements are moving distance learning from the traditional book-based approach pioneered by the Open University in the UK over 30 years ago, to on-line, technology-enhanced courses at all levels.

Globally, increasing amounts of educational and training material are being made available on-line, with a well-publicised example being 'Massive Open On-line Courses' (MOOC). It is widely reported that we are in the midst of a sea change in the way training and education are presented and used (e.g.[5]; [6]). Their supporters regard MOOCs as, potentially, the biggest innovation in education in recent time, with detractors claiming that they can lead to unengaged students, that the difficulty level is not aligned to the skills of the students, assessment is ineffective, and a successful outcome for the student does not lead to a qualification from an accredited course [5].

However, a participation issue with any type of online learning exists for seafarers working in all disciplines; the nature of the work involved requires frequent deployment for extended periods in remote locations. Despite advances in technology, satellite communications bandwidth remains very expensive and thus usually limited to operational transmissions. There is usually very little opportunity for individuals to access the internet through a broadband connection on a regular basis, and sporadic plain text emails, and occasional satellite phone calls are all that are normally available. In response to these issues, and clearly elucidated needs for an accessible and accredited education and training programme from the marine and maritime employment sector, the Marine Learning Alliance (MLA) began developing degrees at Master's and Bachelor's level in 2011, applying technological developments in distance e-learning.

2. DESIGNING AN EFFECTIVE GLOBAL LEARNING PROGRAMME

In order to most effectively develop MLA'S education programmes, an alliance was formed between MLA's academic partner, Plymouth University, a large maritime employer (Fugro), and a professional body (the Institute of Marine Engineering, Science and Technology, the IMarEST). It is worth noting that for many of the sub-disciplines within engineering and wider STEM subjects, students will need exposure to laboratories, equipment or similar to enable them to gain the hands-on skills that their intended profession requires. In such instances, a full distance learning programme cannot provide the entire solution, and, therefore, a blended learning programme is most appropriate. MLA has designed and delivered both full distance learning programmes, and blended learning programmes, and in both examples it was found at an early stage that learning materials have to be designed in such a way so that they are student-centric, with authentic learning using real-world and up-to-date examples. This must feed through to assessments, and it is critically important, particularly with mature learners, to involve them in reflective learning tied to the assessment. This constructivist approach to design yields learners who are aware of what and how they learn, and [4] and [7] place an emphasis on the development of learners to be highly autonomous, self-determined and well-prepared for the complexities of the workplace.

Any higher education programme which is aimed towards a significant cohort of mature learners needs mechanisms in place to recognise and accredit prior learning and experience, and ideally, needs to be as flexible as possible to fit in with a part-time learner's other commitments. Providing tutor support at an individual learner level is another critical aspect of the design stage, which should not be underestimated. Other institutions [8] have experienced issues in the past with designing distance learning programmes which require considerable amounts of tutor support, more in line with the time that would be devoted to a group of residential students. This has led to tensions between management with one eye on the budget, and faculty wishing to deliver an excellent student experience. It is one that can be mitigated to a large degree during the programme design stage, whereby the integration of the student's self-directed learning with a limited amount of tutor support can yield an outstanding learning experience for the student, whilst keeping the cost of delivery of the programme within sensible limits.

One of MLA's key findings in terms of course design, is that taking an existing full-time residential study programme, and turning it into e-learning does not work effectively, whether the learning is aimed at higher education level, or at training course level. Whilst it is widely recognised that identical learning outcomes can be delivered with equal or greater success by a distance e-learning programme, the programme has to be re-designed from the ground up, accounting for the mode of study and the method of delivery. The time and effort involved in this process should not be underestimated by other institutions wishing to develop online or distance learning at Bachelor's and Master's level, particularly if their faculty do not have significant experience in such areas.

Incorporating employer's feedback and input from professional bodies is a critical input throughout the design stage. The role of many professional bodies is not only to be the arbiter of standards within their profession, but is also increasingly seen as informing the content of approved training courses and academic programmes of higher education study. MLA's experience in forming the alliance mentioned above, is that students are significantly reassured by the involvement of a professional body in the course or programme, and it can act to provide a 'standards guarantee' for a new academic programme which has yet to prove itself through many cohorts of graduates entering the employment market. Professional bodies can also help to facilitate employer involvement and industry comment on programme design, ensuring that graduates will be seen as employable and fit for purpose. Many professional bodies have existing employer working groups or 'round table' events, at which course design proposals can be aired and feedback gained, for courses aimed at education or training outcomes.

3. DELIVERING AN EFFECTIVE GLOBAL LEARNING PROGRAMME

Today's student has high expectations of the technology with which they interact. Lectures recorded from the back of the lecture theatre and used again online, simple PDFs placed on a website or any materials where the learner is not immersed and involved will not attract the new breed of technologically literate student, who consider themselves consumers as much as learners. Equally, higher education courses in particular (in comparison with training or CBT-type courses) have to carry a certain amount of gravitas to carry the learner's respect. Whilst this can come from the qualifications of the professor or academic delivering the learning materials, a balance must be struck between using the technology as a facilitator of learning, as opposed to using animations, avatars or similar for the sake of just showing off the technology. Such uses of technology for delivering education, or indeed training programmes associated with a highly-respected and well established professional body may not be appropriate and are likely more appropriately applied to CBT-type scenarios

MLA's programmes, whether fully distance learning or blended learning, are designed to be delivered to a student at a distance using appropriate technologies. One of the key barriers to delivery highlighted above is the nature of the limited access to the internet afforded to much of the seafaring workforce. Evolving technology now allows students to study offline, but immersed within an environment which appears to be fully online, and functions as such from the user's perspective.

MLA offers students the choice of receiving their learning materials in three ways: containing all the learning materials they require to complete a university-level module (between 10 and 30 credits) effectively, they can choose physical media (a USB stick) which is delivered by courier to their home or workplace location, they can download their module from an file sharing location, or they can download the MLA app and content to their mobile device. The first two methods require consumption on a laptop or desktop computer, with the app-based approach allowing the student to consume learning materials on any mobile device, e.g. tablet or smart phone. All three methods present learning materials to the user in very similar, recognisable formats and contain lectures, tutorials, help files (e.g. Helping Engineers Learn Mathematics support), transcripts, assessments, study timetables and course handbooks. Whilst the physical media delivery remains a first choice for some students, it has been overtaken recently, by some margin, by the download or the app choices.

All three methods have been designed using the latest mobile technology to enable the student to download the materials while they have access to the internet, and then consume it whilst at sea with a limited connection. This has been a highly successful element of the MLA's delivery mechanism since 2012, but a rather surprising finding has been that many students who are not seafarers also utilise the offline capability on a regular basis. Many of them consume lectures and tutorials on their mobile devices whilst commuting or during other periods of down time. This evolving 'anytime anywhere' approach to both education and training seems to be a developing trend, which is very much consumer led.

The provision on individual tutor support for each student is one of the key differentiators between MOOCs and other types of technology enhanced distance learning. MLA's learners provide feedback which indicates that 90% see the tutor support they have received during their studies as being critical to their success during their studies. MLA currently has a pass rate of 95%, and module retention and completion rates are also extremely high. Contrasting this with some of the frequently-quoted completion rates for MOOCs (Figure 2), where it is not uncommon for less than 10% of students to finish a module, then it is clear that a level of tutor support can help to reduce feelings of isolation amongst learners, as well as deal with any pastoral or subject-specific issues as soon as they arise. This appears to be a critical element of success for a distance learning programme (whether technology enhanced or not), but effective distance learning tutoring is a skill which is not necessarily within a faculty members' or a trainers' skill set if they are used to traditional classroom-based delivery. Significant training or upskilling within an education or training organisation may be necessary to ensure that support mechanisms meet the expectations of the modern student.

4. CONCLUSIONS, AND LESSONS LEARNED DURING THIS CASE STUDY

There exists a shortfall globally of engineers and technologists, and indeed of those possessed of STEM skills and qualifications within the workforce. Societal and industrial changes, and the rapid pace of technological change means that many Higher Education students are no longer school leavers going to university, but are people upskilling within the workforce, or changing career. Seafarers pose a particular issue due to the nature of their working patterns, the geographically isolated nature of their working environment, and the limits on communication primarily for cost reasons. There exist, therefore, particular issues in the maritime sector in creating opportunities to enhance skill levels through training (primarily for skills development) and provision of education to pursue Bachelor's and Master's degree level study.

Marine Learning Alliance, based in the UK, has worked with a large, international employer, a professional body and a university to design and deliver Bachelor's and Master's degrees, and training courses suitable for Continuous Professional Development, to a global STEM audience. Both blended learning, and fully distance learning courses have been developed, and clear lessons have been learned during this continuing process:

- Care in design is critical
- An existing course which is delivered predominantly in the classroom cannot simply be turned into distance learning or e-learning. It has to be redesigned from the ground up to suit the nature of delivery and the unique requirements of distance learning student, who increasingly have high consumer-driven expectations of their learning experience
- Distance learning is expensive [8], and can lead to tensions between faculty and management, but this can be mitigated by care in the design phase
- Being able to consume, and interact with learning materials offline, is growing in importance, not just for seafarers or the geographically isolated
- Use of the latest technology, and trends in e-learning to meet student expectations is very important, but this should not be the focus of design. MLA has invested in an app-based approach, which works offline. The app allows students (whether on training courses or on university-level programmes) to consume teaching and learning materials, and interact with fellow students on a mobile device. This approach fits well with global trends in mobile device ownership
- Tutoring of students can be time-intensive but is very important in order to maximise student potential as well as retention rates on university-level programmes
- Technology-enhanced distance learning can be highly successful:
    - An alliance can facilitate input from both employers and professional bodies
    - From an institutional perspective, embarking on development of a distance learning programme should not be undertaken lightly, and bringing in an expert provider of technology-enhanced distance learning to work with an organisation with little track record in the area can yield significant economies. It can take enormous investment to design, accredit, prepare and deliver a distance learning programme, whether at Bachelor's, Master's level or for a training course. Experience can help significantly, and MLA's experience shows that it is now possible to prepare and deliver a 100 hour, 10 academic credit academic module in 10% of the time taken only 4 years ago, using institutional experience within it's expert team of academics, learning technologists and student support specialists.

The modern student, whether studying on a professional-body endorsed training course or on a full university level Master's degree has very high expectations of their student experience. With rising tuition fee levels, students are behaving much more like customers purchasing luxury goods and services. They expect outstanding quality from their tuition, support and learning materials. These three elements can be provided by technology-enhanced distance learning, which tends to fit extremely well with modern working patterns and practices. Alliances, such as the Marine Learning Alliance highlighted in this paper; and which can include a specialist distance learning organisation, a professional body, an employer and a university, yield significant financial economies, improve traditional course design, reduce time to market, and ultimately provide the consumer-driven student with an outstanding learning experience.

5. REFERENCES

1. Engineering UK (2015). The state of Engineering. Engineering UK, London. Available at http://www.engineeringuk.com/Research/Synopsis/ [Accessed 9 August 2016]

2. Manpower Group (2015). 10th Annual Talent Shortage Survey. Available at: http://www.manpowergroup.com/ [Accessed 8 August 2016]

3. Department for Transport, 2015. Maritime Growth Study: keeping the UK competitive in a global market. 134 pp.

4. KETTLE, J. 2013. Flexible Pedagogies: employer engagement and work-based learning. *The Higher Education Academy*, York. 43pp.

5. FREITAS, S. I., J. MORGAN AND D. GIBSON. 2015. Will MOOCs transform learning and teaching in higher education? Engagement and course retention in online learning provision. *British Journal of Educational Technology*, 46, 3. 455-471.

6. KURZMAN, P. A. (2013) The Evolution of Distance Learning and Online Education, *Journal of Teaching in Social Work*, 33:4-5, 331-338, DOI: 10.1080/08841233.2013.843346

7. BLASCHKE, L.M. (2012) Heutagogy and Lifelong Learning: a Review of Heutagogical Practice and Self-Determined Learning. *The*

International Review of Research in Open and Distance Learning. 13 (1), 56-71.

8. DOIG, A and C. PEKCAN. 2015. An Online Master's Degree: teaching and learning strategy vs. managerial constraints. *ICICTE 2015 Proceedings* 192-201.

6. AUTHOR'S BIOGRAPHIES

Dr. Richard Thain is Academic Director and Chief Executive at the Marine Learning Alliance. Dr Thain has a PhD in physical oceanography, and a postgraduate qualification in Learning and Teaching in Higher Education. He is a Fellow of the IMarEST and of the Higher Education Academy, and has been designing and delivering successful technology-enhanced distance learning programmes since 2011.

Dr. Jaimie Cross works as a Lecturer in Marine Science at the Marine Learning Alliance. Dr. Cross completed his PhD from Plymouth University in 2012, specialising in the fields of Physical Oceanography and Sediment Transport dynamics. He is an experienced teacher and communicator of marine science, and is the author of several peer-reviewed journal articles.

David Kelly is the Director AsiaPac at the Institute of Marine Engineering, Science and Technology (IMarEST). Mr Kelly is a Chartered Marketer based in Singapore, and works throughout the AsiaPac region developing professional body, education and training opportunities.

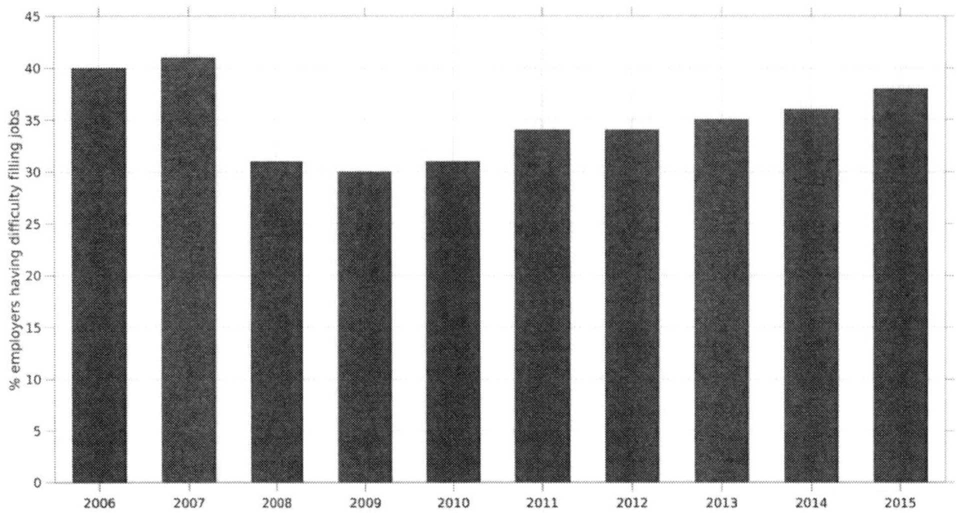

Figure 1: Percentage of employers surveyed who express difficulty in attracting candidates to fill vacant positions across the period 2006 to 2015. Of these, for the period 2011-2015 three of the top five most difficult positions to fill were skilled trades, engineers and technicians respectively (data from [2])

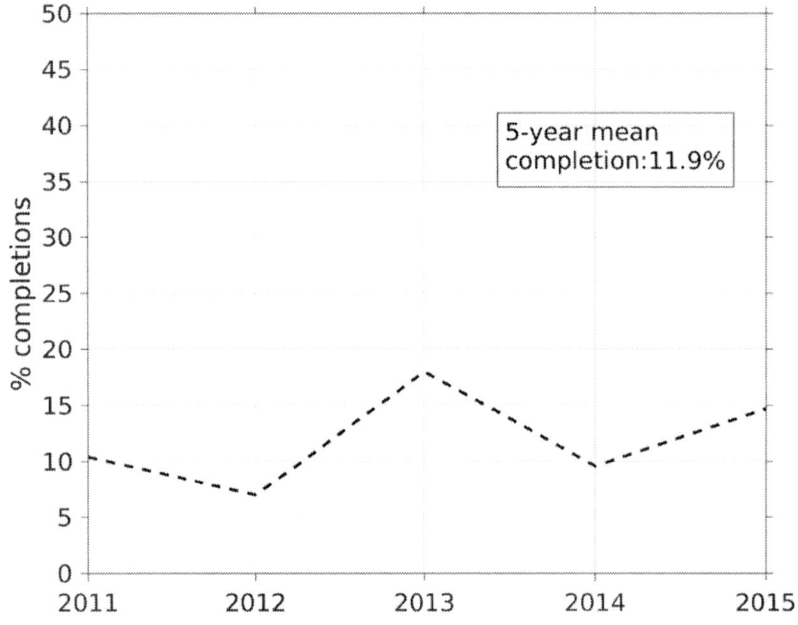

Figure 2: Completion rate for Massive open online courses (MOOCs) for the period 2011-2015.
Data acquired from Jordan (2016 available at http://www.katyjordan.com/MOOCproject.html) for selected MOOCs from a range of global providers. The mean completion rate for these courses across this period is as displayed.

TECHNOLOGY, COLLABORATION, AND THE FUTURE OF MARITIME EDUCATION

R Burke and **C Clott,** Maritime College, State University of New York, USA

SUMMARY

This paper seeks to explore the dimensions of a strategy for integrated training and education of the maritime professionals of the future. Maritime academies worldwide face the task of training ships' officers by employing different models. Although the American model that is university based treats the training of mariners in a manner similar to professional education in fields such as engineering and management, aspects of the STCW Code rest upon the longstanding apprenticeship model used in Europe and Asia. The American experience in the years after World War II was that experiential training would not adequately prepare individuals to perform in an environment of rapidly developing technology. Faced with enhanced use of information technology in ship operations with more remote and autonomous operation of ships and ports, what are the implications faced in employing various educational models for training mariners, and what will be the interactions of collaborative activities with other aspects of technical and managerial education of transportation professionals?

1. INTRODUCTION

The maritime industry is undergoing technological change at an unprecedented rate. The sizes and speeds of merchant ships have increased dramatically, as have the variety and volume of cargoes. Information technology efficiencies have driven the industry, ashore and afloat, from semi-automation toward full-fledged autonomous operation. Instantaneous, world-wide communication has transformed the ship's "Master under God" into the Operations Manager of a multi-million dollar corporate asset. Decision-making has become increasingly centralized on shore for all elements of the voyage. Age-old training in Marlinspike seamanship, chipping and painting, celestial navigation, and lagging of pipes are among the skills with rapidly diminishing importance.

At the same time, the volume and variety of international regulation has exploded. Once limited to safety of life and public health, there are now multiple international agreements that address such issues as energy efficiency of ships, the potential for both water and air pollution, the weights of loaded containers, the safety of sea life, and crew training and certification. Every year the regulatory burden grows for ships.

This is the environment in which we must examine the future of maritime education and training. Major changes will affect every aspect of marine transportation.

1.1 DIMENSIONS OF MARITIME TRAINING & EDUCATION

The ship is but a single link in the supply chain, but multiple links of that chain are part of the maritime world. To address maritime education and training we must consider all of the relevant dimensions. These include:

- Transportation economics
- Shipping finance
- Ship design, construction, and repair
- Ship crewing, operation, and agency
- Port design, construction, and operation
- Freight sales, consolidation, and forwarding
- Customs documentation and administration
- Intermodal coordination

This list, undoubtedly incomplete, covers a breadth of disciplines including business and finance, engineering, law and marine science. This is in addition to the specific technical knowledge of navigation, ship stability, trucking, rail, power plants, and labor relations that must mariners must be competent in. This breadth of knowledge is unfortunately not taught to prospective mariners. Current licensing requirements reward specialists who lack an understanding of the multiple challenges faced in this demanding occupation.

1.2 A HISTORICAL PERSPECTIVE FROM THE UNITED STATES

Historically, educated people did not seek careers as mariners. The work was hard and dangerous, and formal schooling offered no advantage. Formal education of mariners was finally instituted by the establishment of the U.S. Naval Academy in 1845 and the founding of a number a state and local maritime academies beginning in 1874. Motivated by the frequency of maritime disasters resulting from perceived mariner incompetence the training was formal and more practical than the present and did not confer an academic degree. Then as now, significant periods of sea service were expected.

In the 20th century following World War II, American maritime academies became degree-granting institutions associated with larger state university systems, eventually offering broadened educational degree programs that address all of the dimensions of maritime training in addition to courses normally associated with higher education. Business courses were added to maritime curricula to provide a greater education in the shore side aspects typically performed by qualified marine officers at ports of call. As STCW requirements have been continually upgraded there is less room for routine business courses in subjects such as accounting, marketing, administration and finance.

Business, among all occupations, has been experientially based throughout history although business administration topics such as accounting, marketing, administration, and finance have become the basis of extensive undergraduate education only since the early 20th century. One in five of American undergraduates are engaged in some form of business education. The growth of graduate study in this area has been styled as "management education," and has become focused on many of the more technical aspects of management practice and strategy. The premier management degree is the MBA, a subject of some modern controversy, which Mintzberg [1] called "...a 1908 degree based on a 1950s strategy."

The United States suffered from a lack of trained engineers for more than a century after its independence, as evidenced by struggles with major technical projects such as the construction of the Erie Canal [2]. Along with other disciplines, degree programs in naval architecture and marine engineering blossomed after the 1862 passage of the Morrill Act, which established "land grant colleges" in the various states to promote "agriculture and the mechanic arts." Engineering education in the United States was nevertheless very practically oriented, and career paths after graduation depended heavily on experiential exposure.

Real change to the structure of engineering education did not come until after World War II, an evolution recounted by Hazelrigg [3]. The war had brought rapid technological development in electronics, nuclear energy, and aviation, and it had been discovered that engineers brought up in an experiential, apprentice-style educational system were unable to contribute as needed when technology changed rapidly. The response, planned before the war but implemented first at the Massachusetts Institute of Technology after the war had ended, was to make engineering curricula deeper in mathematics, the sciences, and analytical engineering subjects. The perspective was that the graduates would develop applications experience on the job after graduation, but would retain the basic understanding necessary to develop new technology from scratch. Research was a major activity in this scheme. This trend was steepened by the post-Sputnik focus on mathematics and science; the result was a half-century of rapid technological development.

The trend toward an analytical education for engineers in the U.S. has been relaxed in recent years. The Franklin Olin College of Engineering in Massachusetts was founded in 1997 to radically change engineering education by emphasizing creativity and experiential learning. The president of Olin, R.T. Miller, was quoted as saying [4]

"At the core of the problem was the fact that engineering students were not very engaged in their education... Most of the courses had to do with analysis and calculus, with the unspoken assumption that more math is always better... First, we defined the term: an engineer, in our vision, is a person who envisions what has never been, and does whatever it takes to make it happen. The science is just a set of power tools that enable engineers to make it happen; they are not fundamentally what an engineer is or does."

1.3 INTERNATIONAL REGULATION OF MARITIME EDUCATION & TRAINING

With respect to qualification of mariners, two principal regulatory stems branch from the work of the United Nations: the International Maritime Organization (*IMO)*, and the International Labor Organization (*ILO)*. IMO also is the principal international regulatory agency with respect to matters of ship design, construction, and operation.

From a mariner's perspective, their qualification requirements are largely governed by IMO's *International Convention on Standards of Training, Certification and Watchkeeping for Seafarers, 1978*, as amended [5] (hereinafter *STCW)*. The stated purpose of this document is "...to promote safety of life and property at sea and the protection of the marine environment by establishing in common agreement international standards of training, certification and watchkeeping for seafarers." Prior to this agreement there was significant public concern regarding global variations in mariners' qualifications, especially since it was often rumored that certificates easily could be purchased in the black market. The establishment of an international standard was also intended to facilitate the employment of mariners from diverse locations. STCW imposes a skill-based training framework upon mariner qualification, with deep roots in the apprenticeship model. This is not surprising; many of the mariners directly engaged in the drafting of the convention were products of that system.

Unfortunately, that system is not used everywhere in the world, and it imposes severe constraints on education and training systems that differ. The United States, among other countries, applies a more collegiate-centered approach to such training. The apprentice-based, skills-oriented approach is difficult to apply in such an environment.

The implementation of the code by the U.S. Coast Guard in the United States is focused on distinct and measureable achievements, i.e., for engineers burning a 75 mm round hole in steel plate, and attending lectures with a specified amount of "seat time." In addition to classroom and skill training, the code requires at least a year of sea service before issuance of a certificate; by comparison, it is possible to obtain an aircraft commercial pilot's certificate with eighteen weeks of training.

While STCW deals with the responsibilities of mariners to become qualified, the *Maritime Labor Convention of 2006* (hereinafter MLC), established through the ILO, defines the responsibilities of maritime employers to mariners [6]. The articles were set forth because the convention was "Desiring to create a single, coherent instrument embodying as far as possible all up-to-date standards of existing international maritime labour Conventions and Recommendations..." Recent amendments, which may come into force soon, are aimed at eliminating bullying, harassment, forced labor, and documentation.

The international regulatory climate for ship operators, naval architects, marine engineers, freight forwarders, and other parties employed within the supply chain, has become significantly more complex recently. Aside from design requirements established for safety of life at sea and prevention of oil pollution, all manner of topics addressing commerce, the environment, and technical performance have become the focus of new regulations and recommendations that affect both design and operation of the entire supply chain. Technical or commercial prowess on the part of those who work in the supply chain is no longer sufficient; now, they must be well versed in the all of the aspects of international technology, commerce, and law.

2. A PRESENT AND FUTURE VIEW

The maritime industry is a notoriously conservative one and lags as early adopters of new technologies; valuing cost containment and reliability above all else. At least a portion of this reluctance has to do with the cost and difficulty of hiring well-qualified crews. New technologies inevitably come to the commercial marine world from the military.

The world around us, however, is on a steep upward technological path. Terms like big data, artificial intelligence, autonomous vehicles, augmented reality, and wireless communication are no longer restricted to academic journals; rather, they are characteristics of our world and of the products that we buy and use.

We know that not all of those who speak of the future are prophets, and that even those who are prophets may not be welcomed. Nevertheless, we see five major trends occurring. They are, in no special order:

- *The diminishing importance of crews on modern ships*, brought about by the reduction of crew sizes as ships have increased in size. The master and mates have the principal ship operation responsibilities, while engineers do very little operational work outside of maneuvering and emergencies. Important tasks such as planning cargo stowage, ordering supplies and spares, and determination of stability are done ashore. Outside of housekeeping and preventive engineering maintenance programs, almost no traditional shipboard maintenance is done by the crew. This is not to say that the need for integrity, responsibility, and positive action from the crew aboard modern ships is diminished; in fact, it is probably enhanced by the fact that such demands are shared among fewer people, and that there is more at stake. One could argue that many of the recent major casualties were not technical failures as much as human failures resulting from crew fatigue, lack of responsibility, or poor communication.
- *The need for detailed data from all elements of the supply chain, easily collected, transmitted, and shared, remains undiminished, growing in importance and volume.* This is the key enabler of the tremendous recent growth in the volume of world trade. The interdependent coordination of the various elements of the supply chain ashore, i.e., trucking, rail, pipelines, etc., at sea, i.e., cargo ships and support vessels, and at the interface, meaning terminals, marshaling yards, stevedoring, and forwarding, is entirely driven by data. The more effective the use of data, the more efficient then is the coordination of the supply chain; however, cyber threats and a corresponding need for security will continue to constrain this development.
- *Growth of the body of international regulation, both in depth and extent*, which is likely to include regulation of ship operation in far more detail than at present. This is likely to include the establishment of specific operating routes and corresponding traffic control, much as exists currently in the commercial aviation. Given that ships are now routed for fuel economy and that port congestion is a significant problem, it is natural to believe that departures, routing, and arrivals will be regulated in the future. Also, the role of human factors, especially fatigue, communication, the delegation of authority will likely attract closer scrutiny.
- *The evolution of ship automation into autonomous operation with centralized control*, developing much as we are seeing the aviation world develop. This idea is treated with some derision; however, that was also true with respect to autonomous aircraft operation. Less than thirty years ago, a somewhat famous (or infamous) news report [7] stated that a University of Miami professor of management science in jest told the Airline Pilots Association about 21st century aircraft by saying 'The crew will consist of one pilot and a dog. The pilot will nurture and feed the dog. The dog will be there to bite the pilot if he touches anything.' Jokes notwithstanding, at the time of that publication the operationalization of military autonomous aircraft was already underway, and now we are on the verge of seeing autonomous commercial aircraft go into service. As for the maritime industry, an excellent report by Lloyds Register et al [8] foretells of higher levels of autonomy, especially among navies, with technological follow-up in the commercial maritime world. In fact, the

US Navy has built and is testing an autonomous surface combatant [9]. These technologies also depend heavily upon computers and networking; thus, we face an age where all control systems will be based on TCP/IP networks and the technologies that follows them. In terms of operational training, an article in *The Economist* [10] gives anecdotal evidence from the aviation industry that operators' paradigms must be adjusted for autonomous or near-autonomous operation, which suggests that a reconsideration of the STCW framework must follow.

- *A fading away of seafaring as a profession, together with mariners' traditions and practices that have developed over the millennia.* We have many colleagues and friends who consider this to be unthinkable, but the technological tide seems to have turned. First, the ship has become just a single element of the supply chain, having far less value than the cargo carried or the shoreside infrastructure with which they interface. Secondly, ships have become less anthropomorphically personalized, which is to say that they are just standardized products and capital assets. Lastly, enhanced automatic control tending toward autonomous operation further reduces the need for crew. Among the motives are widely reported crewing shortages in recent years. This will not be improved as the required level of technical expertise increases. Regardless of modern conveniences, a career at sea is not the most desirable life-style. The work is still relatively demanding and dangerous, and it is not conducive to family life. All in all, if we maintain the status quo we can expect crew shortages to be exacerbated.

3. SOME MODEST PROPOSALS: BREADTH, COLLABORATION, AND INTENSITY

3.1 BREADTH

Given the historic lessons in the aftermath of World War II for engineering, we suggest that maritime technology is changing today faster than ever with the promise of future acceleration. We need to radically broaden and deepen basic education for engineers, operators, and those entering the business functions. This should include:

- Significant education in information technology with an emphasis on cyber security for ship designers, ship operators, all shoreside personnel.
- An understanding of economic geography, including aspects of navigation and climatology for all ship designers and for all shoreside personnel, as well as the technical aspects of each mode of transportation as they relate to economic and geographical constraints.
- Education of those who will serve in business functions in the maritime industry must, first of all, have a deep background in business. The five major aspects of a good business education – marketing, operations management, economics and financial management, organizational behaviour and personnel management, and information systems management – cannot be neglected for the sake of specialized education pertaining only the maritime link in the supply chain.
- Make business more a part of a technical education, and technology more a part of a business education. The complexities at the intermodal interfaces now have a huge impact on the effectiveness of designs. Just as naval architects must know the complexities of the entire supply chain, those working ashore in the supply chain must have basic understanding of the technology of ships.
- The people who serve on the relevant IMO committees are intelligent, dedicated professionals; nonetheless, they cannot escape their own experiences. Too many of the delegates engaged in matters covered by the STCW convention unfortunately have experiences that reflect a narrow operational point of view rooted in past practices. While one may be proud of a successful sea career culminating in service as a ship's master, new people with newer experiences must be brought into the process. Tradition can be a sea anchor in a storm, or it can be just a drag. We also must broaden the function of international regulation.

3.2 COLLABORATION

Many maritime educators work at specialized institutions. We too have our traditions and are slaves to our own experiences. If we are to achieve breadth, we must seek collaborative alliances with other institutions, with institutions that have broader views and a more progressive outlook. We have been too slow to accept new developments in technology, and it has been easy to hide from it at specialized maritime institutions.

Collaboration requires that we admit to ourselves that although we are different, we are not *that* different from other institutions.

3.3 INTENSITY

In our opinion, the training of shipboard operators has taken a wrong turn by focusing on sea service time. Anyone who has made a trans-oceanic voyage understands that it is long, and often boring. The training that results from such exposure is by definition *low intensity*, in that significant learning experiences generally between long intervals.

How can the aviation industry train someone to the level of a commercial pilot's certificate in eighteen weeks? It is achieved by *high intensity* training. Such training is easily done using high quality computer-based simulation, and a syllabus that builds situational awareness by exposing students to repeated crisis-response scenarios of progressively increasing levels of risk and complexity. Such simulators are available for training shipboard mates and engineers, but the use of them has not been generally seen as an effective replacement for sea service. We should be able to take an individual with an otherwise appropriate technical education and train that individual to a satisfactory level of competence within a similar period of time *if the training is of high intensity*.

Another aspect of enhanced simulator training is that the simulator will probably bear a distinct similarity to the type of equipment that will be used to operate vessels remotely. Is simulator training a pre-cursor to the kind of training and watchkeeping that the maritime industry can expect in the future? Perhaps the future is right here in front of us.

4. CONCLUSIONS

The authors contend herein that the education and training for the maritime industry requires significant modification in the face of rapid technological and organizational change. We believe that the efforts required include a broadening of the educations that are received by all in the industry, by achieving this broadening through collaboration with broad-based educational institutions, and increasing the intensity of the operational training offered to those who have received. The traditions of the maritime industry, the traditional roles of those working in the industry, and means by which they learn those roles will be changed dramatically.

5. ACKNOWLEDGEMENTS

The authors are grateful to ABS and the Maritime College of the State University of New York for their support; however, the views presented herein are the authors' alone and do not represent those of others.

6. REFERENCES

1. MINTZBERG, H., *Managers Not MBAs: A Hard Look at the Soft Practice of Managing and Management Development*, Berrett-Koehler Publishers, 2004

2. KELLY, J., *Heaven's Ditch: God, Gold, and Murder on the Erie Canal,* St. Martin's Press, 2016

3. HAZELRIGG, G., *Systems Engineering: An Introduction to Information-Based Design*, Prentice Hall, 1996

4. EUCHNER, J., "The Future of Engineering Education: An Interview with Rick Miller," *Research-Technology Management*, 2014

5. http://www.imo.org/en/OurWork/HumanElement/TrainingCertification/Pages/STCW-Convention.aspx, 2016

6. http://www.ilo.org/global/standards/maritime-labour-convention/lang--en/index.htm, 2016

7. Fortune Magazine, Sept. 26, 1988

8. *Global Maritime Technology Trends 2030,* Lloyd's Register, QinetiQ, and University of Southampton, 2015

9. "DARPA Launches Autonomous Sub Tracker," Maritime Executive, March 31, 2016

10. "Cockpit automation - Difference Engine: Crash program?" *The Economist,* August 26, 2013

7. AUTHORS BIOGRAPHY

Richard Burke, Ph.D., FRINA, holds the position of ABS Professor of Naval Architecture and Marine Engineering at Maritime College, State University of New York, USA. He is responsible for teaching and research in the College's academic engineering programs. His previous experience includes twelve years directing engineering programs at Maritime College, a sabbatical year appointment as Presidential Fellow at the Research Foundation of the State University of New York, a stint as a management faculty member at Rensselaer Polytechnic Institute, and more than fifteen years of industrial experience.

Christopher Clott, Ph.D., holds the current position of ABS Professor of Marine Transportation and Logistics at Maritime College, State University of New York, USA. He is responsible for teaching and research in the College's academic programs in international trade and transportation management. His previous experience includes work in ocean transportation with OOCL and Lykes Lines and administration and faculty roles at the California Maritime Academy, Saint Xavier University, Elmhurst College and the University of St. Francis in the U.S. His recent research has focused on shipping alliances, port congestion, intermodal freight transportation and employment.

A STUDY ON THE DEVELOPMENT AND IMPLEMENTATION OF MASSIVE OPEN ONLINE COURSES IN MARITIME EDUCATION AND TRAINING

I F Muhammad and W E Yutuc, Malaysian Maritime Academy (ALAM), Malaysia

SUMMARY

Due to technological advances, global issues and new regulations in the shipping industry, the seafarers are tasked to constantly remain current and comply with the required industry standards by ensuring their competencies are updated. For the purpose of creating another platform for the seafarers to meet and accomplish this requirement, this paper studied the development and implementation of Massive Open Online Courses (MOOCs) in Maritime Education and Training (MET). The approach considered instructional design strategies (ISD) and methodology, target audience, type of training, content, evaluation and assessment schemes. Furthermore, the significant barriers in MET have been discussed including the current issues and challenges faced in the maritime industry. Given these complexities for MOOCs and considering the expected outcomes for both the seafarers and the maritime industry as a whole, the process may seem wearisome. Also, its immediate implementation may negatively impact the expected performance outcomes of the entire system as it currently stands. However, the long term benefits of a properly structured MOOC, being a learning management system (LMS) or e-learning platform, should be seen to overcome this setback and serve its intended purpose effectively. This paper presented a MOOC model specifically designed to conform with the requirements of the International Convention on Standards of Training, Certification and Watchkeeping for Seafarers Manila Amendment (STCW 2010).

1. INTRODUCTION

Technological advances, global issues and regulations demand change and create significant momentum and impact on the way seafarers interact with the equipment onboard the ships where they work. For instance, seafarers working onboard modern ships with unmanned machinery spaces (UMS) and distributed control systems (DCS) do much of the monitoring and controlling of shipboard operations with the use of computers. This is far different compared to the way seafarers did their job before computer technology was widely used onboard ship. To adapt and cope with these changes and developments, the seafarers are tasked to constantly remain current and comply with the required industry standards by ensuring their competencies are updated. For the purpose of creating another platform for the seafarers to meet and accomplish this requirement, this paper studied the development and implementation of Massive Open Online Courses (MOOCs) in Maritime Education and Training (MET). It considered the capability of online vehicles like e-learning, student and school management systems, virtual classrooms and learning management systems (LMS) to address competency issues.

In this paper, the instructional design strategies (ISD) and methodology, target audience, type of training, content, evaluation and assessment schemes were considered in developing online courses. It was also aimed to contribute to the goals of the International Maritime Organization (IMO) on safe, secure and energy efficient shipping operations on cleaner oceans [1]. The MOOC model presented was specifically designed to conform with the requirements of the International Convention on Standards of Training, Certification and Watchkeeping for Seafarers Manila Amendment (STCW 2010). Furthermore, the significant barriers in MET have been discussed including the current issues and challenges faced in the maritime industry.

Although there is a possibility that the immediate initial implementation of MOOCs in MET may negatively impact the expected performance outcomes of the entire system as it currently stands because of leadership challenges, misalignment of goals and priorities, competency, and change management issues; the long term benefits of a properly structured and implemented MOOC, being an LMS or e-learning platform should be seen to overcome this setback and serve its intended purpose effectively.

2. MOOCs IN MET

Many innovative e-learning technologies are in use today. MOOCs is one, conducted on a massive scale with unlimited participation of users with an open access via the web, as the acronym implies. It is also regarded as a non-defined pedagogical format to organize learning, teaching and training on a specific topic in an informal, online and collaborative way [2]. In the process of its evolution, MOOCs was further classified into two distinct types: the "cMOOC" and the "xMOOC". The cMOOCs are primarily based on connectivist pedagogy while the xMOOCs have a much more traditional course structure typically with a clearly specified syllabus of recorded lectures and self-test problems [3]. In the MET environment, MOOC utilization is not yet that prominent, although it is believed that its capabilities can address both learning and competency outcomes required by the stakeholders and the industry [4]. A homepage screenshot of a MOOC running on a Moodle LMS platform [5] is shown in Figure 1.

Leveraging on the emerging technologies in the form of e-learning as applied to MET, a "maritime education

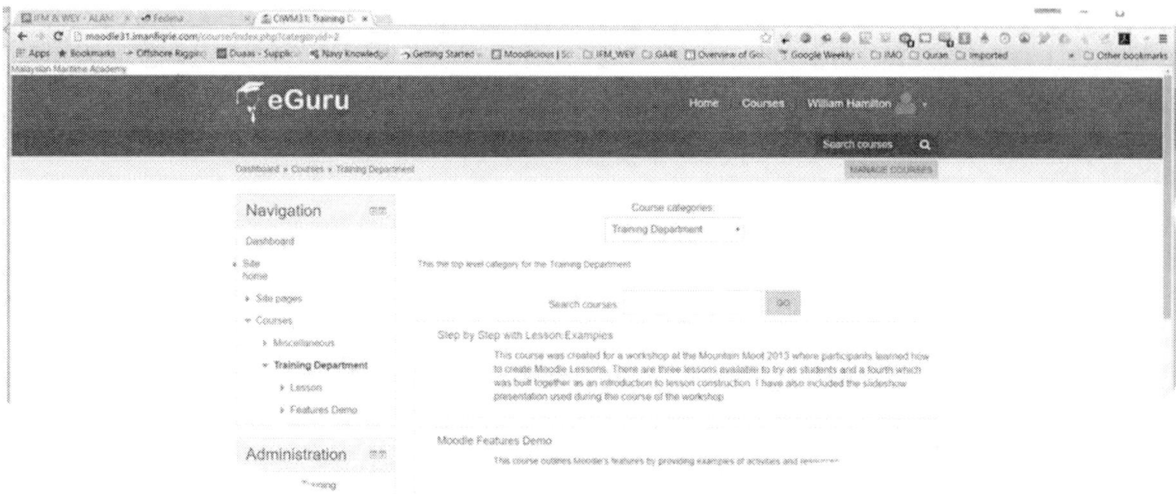

Figure 1: Sample MOOC on a Moodle LMS platform [5].

model" then, should be one that includes competency based outcomes with mechanisms that will enable the transfer of these "blended competencies" to affect business results and effectively combine instructional quality, lower costs, and increased access through unlimited scalability, which are all potentially doable using MOOCs. An example of this learning competence for learning professionals is the ATD Competency Model shown in Figure 2.

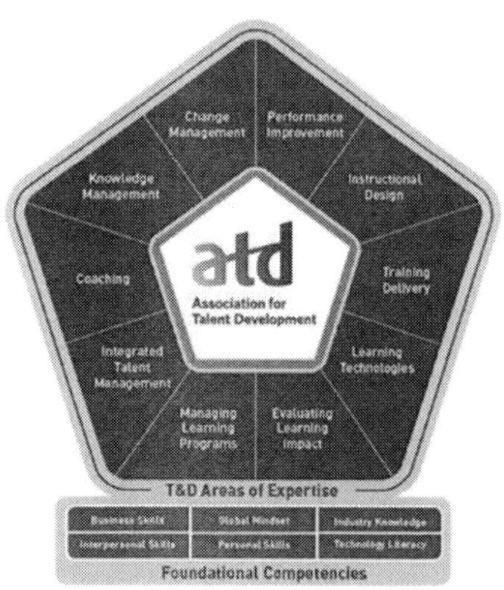

Figure 2: ATD Competency Model [6].

Essentially, the ATD model demands that a certified training or learning professional have both foundational competencies and 10 areas of expertise that includes such areas as performance management skills, instructional design skills, training delivery skills, learning technologies skills and several other area skills including some foundational skills like global mindset and technical literacy to name a few [6].

According to Trachtenberg [7], as early as 2012, MOOCs have been a part of "a seismic shift in online learning that is reshaping higher education". The same author goes on to suggest that MOOCs were gaining international recognition by some well-known universities as early as 2011. Nowadays, it is possible to host an entire digital or online student and school management system (essentially an e-Digital University) in the cloud, as shown in Figure 3. What is even more remarkable is that these applications can now integrate with a LMS like Moodle.

As to courses and MOOCs, several other differences still exist between stand-alone courses, MOOCs and organized MOOCs. MOOCs are becoming more and more sophisticated and more and more relevant for both professional and career development. The authors believe that curriculum-based MOOCs will drive innovation as it will be a significant disruptor for the paid model of traditional education systems. As MOOCs have more context than stand-alone courses, they are also being recognized by institutions such as Colorado State University's Global Campus, Antioch University, San Jose State University and Georgia State University as credit for knowledge, skills and abilities [9]. Also the traditional role of the lecturer, facilitator and student is not a given or necessarily significant in this forum and learning can happen both ways in these kinds of connectivist networks in lieu of the traditional closed classroom style of learning. In this digital environment, there are fewer constraints than that of the classroom. A number of governments recognize this and have specific programs for the development of these 21st Century Skills that involve increasing digital access and minimum IT skills. For instance, the U.K. utilizes e-Portfolios and Pebble PADs [10]; the U.S. employs the 21st Century Learning education curriculums involving collaboration, communication, global awareness and information literacy [11]; and Malaysia calls out for 21st Century Skills involving

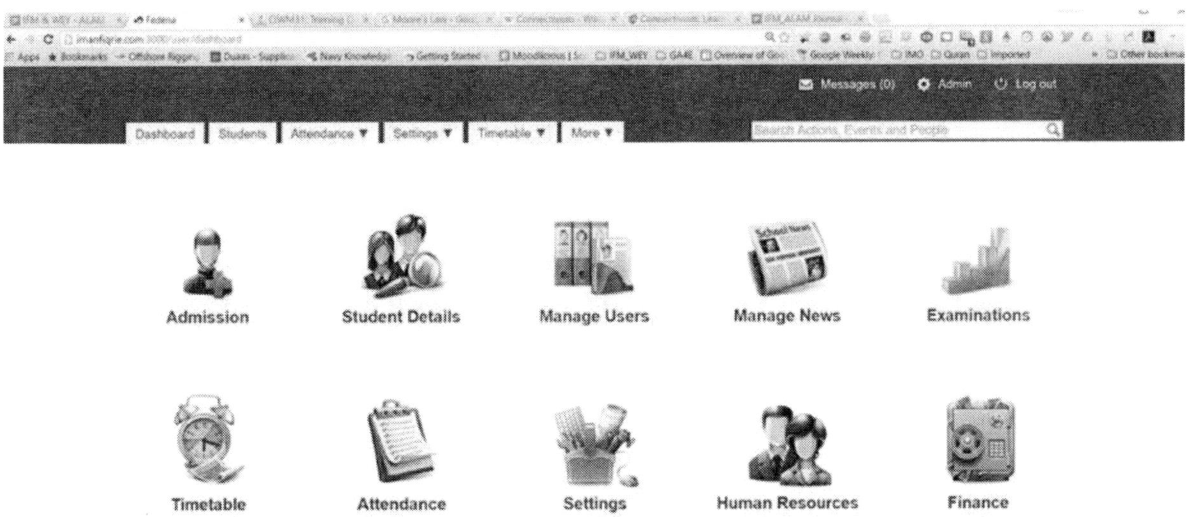

Figure 3: A sample Fedena Community Project Student and School Management System [8].

increased digital access, Smart Schools, IT infrastructure and leveraging ICT for learning in their Malaysian Education Blueprint 2013-2025 [12].

MOOCs also enable the development of Personal Learning Networks (PLNs) or Communities of Practice (COP). Since it has been noted that over 80% of potential knowledge comes from informal learning [6] both PLNs and COPs are a fantastic way to take advantage of this. A good example of a PLN vehicle would be Yammer. It's like a cross between Facebook, chat and email. Microlearning (very short content) can easily be done. An example of COPs where people who share common interests can be seen on the ATD website (https://www.td.org/).

3. MOOCs AND STCW 2010

In the development and implementation of MOOCs in MET, over focus on LMS without a thorough look at STCW, its requirements and industry would be a grave mistake in achieving the overall goals previously discussed. It is important to note that being one of the four pillars of the International Maritime Organization (IMO), STCW sets the qualification standards for officers and watch personnel on seagoing merchant ships, including the competency standards in line with IMO's goal to maintain a safe, secure, energy efficient ship on cleaner seas. In 2010, STCW adopted a new set of amendments in Manila called "The Manila Amendments" or STCW 2010. Some significant amendments include certain training standards in line with new technological and operational competencies required to perform the seafarer's duties onboard. The amendments to STCW 2010 were made effective 1 January 2012 with a transition period until 2017 wherein all seafarers must be certified and trained according to the new standards [13].

Many of the aforementioned concerns cannot be addressed in this one paper as the central thrust is creating another platform (physical and theoretical) to help seafarers meet the multitude of requirements and regulations. The platform studied here was an LMS, specifically MOOC. A search of the scholarly literature for materials on the subject of development and implementation of MOOCs in MET seems dismal at best. On the web, however, seems subject matter experts and content abound and more accommodating in the way of materials on the subject; the technology and content are fluid on the web and leads "scholarly" material.

Nonetheless, a search for the top five online maritime courses providers yields Llyod's Maritime Academy, Videotel, Llyod's Register, Coracle Online and Shipgaz as some of the best at offering online training or courses [14]. The pattern that seems to develop is that a lot of customized courses that seafarers could use is presented, but seems to be void in the STCW courses and content. Thus the title of provider of online maritime courses appears to have a caveat, e.g., provider of some maritime courses.

Again, if the objectives are to be driven by the aforementioned IMO outcomes, then those courses seem to be not available online. This is a problem, again, if the stated objectives of using an LMS is to help the seafarer meet and accomplish the many requirements as stated, one must conclude that a large performance gap exists.

As to the STCW and assessed needs, it is not the purview of this particular study to challenge them outright. However, a look at many maritime outcomes that fit the categories exposed by the IMO as international maritime goals yield a solemn picture of catastrophes, unsafe seas and acts as witnessed by fires, collisions, pollution, piracy and conditions of vessels on the high seas [15].

Again, the assessed needs of the IMO seem far apart from the reality of industry. Having said this, STCW 2010 is what we have and must, therefore, be the instrument with which to measure any curriculum, course, LMS, MOOC or outcomes.

Additionally, after needs, objectives and outcomes have been formulated, one of the first orders of business should be the creation of the evaluation, assessment tools and mechanisms prior to any content creation, whether online or not. This kind of process does not appear readily visible, mainstream or the order of the day. However, a MOOC/LMS with good organizational processes is seen to achieve the stated objectives of helping mariners meet their competency objectives.

4. MOOCs DESIGN AND DEVELOPMENT FOR MET

In terms of design, MOOCs are still considered in their early stages of development and are still evolving. If one looked at the business product cycle graph to draw a comparison, where in the cycle would the "MOOC product" be (Introduction, Growth, Maturity and Decline), it may be suggested that the Introduction Stage as profits are low to non-existent, research and development plenty.

At Stanford University, MOOCs were initially developed by the professors, and later on by MIT and Harvard instructors. They were based primarily on a strongly behaviourist, information transmission model, the core teaching being through online recorded videos of short lectures, combined with computer automated testing, and sometimes through the use of peer assessments. They were delivered through special cloud-based software platforms such as Coursera, Udacity and edX [16]. In general, as can be applied to the MET environment, the xMOOCs design features are shown in Figure 4.

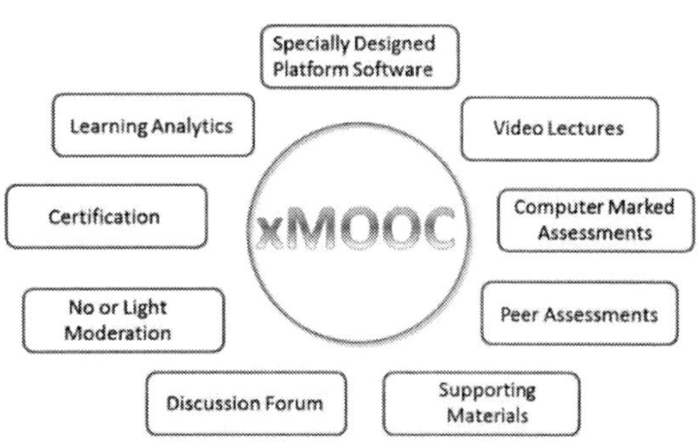

Figure 4: xMOOCs features used in early cloud based systems.

When applied to the study to conform with the requirements of STCW, Figure 5 below clearly emerges an evolved model that can be used in MET. The standards demanded by a senior maritime officer were that the same requirements, curriculum, content, assessment, etc., would have to be accomplished online. Figure 5 does that quite nicely and gives clear alignment of such between MET course methodology and that used by the aforementioned xMOOC designs.

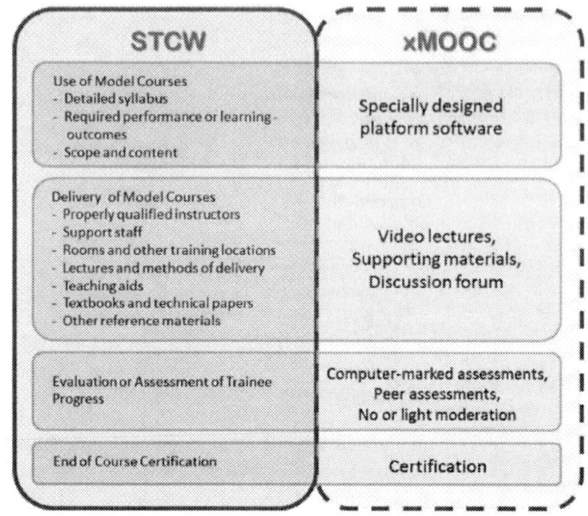

Figure 5: STCW - xMOOC featured course design methodology.

Issues on what can be realistically covered online, what must be "blended" off-line and how it might be done must be addressed if there is to be acceptance by state maritime authorities of any model for online courses or MOOCs.

In MOOCs design, the talent or training professionals, must among other things, know the tools of the trade and know how to apply them. In the case of transforming content, they should have analytical knowledge of instructional design (ISD) as this is paramount to the process. ISD is one of the ATD Competency Model competencies previously mentioned. ADDIE is more than one such ISD model, it is the 'gold standard' of ISD. Others exist like SAM (Successive Approximation Model) and Rapid ISD (ADDIE cut short), to name a few.

A thorough evaluation phase is required of ADDIE, as represented by the last acronym 'E'. If this is cut short, 'real' transformation in the maritime industry and MET is at risk. Firstly, the MET solution process is not just about providing training for some perceived objectives. According to the Association of Talent Development (ATD), performance issues occur at the organizational, process and job level [6]. Typically, most jump right at the job or individual level as the solution where training is the one which needs to be fixed. Giving little time for proper assessment of needs that may reveal the organization or process as 'the problem'. More to the point, an ISD methodology facilitates transformation as it allows for the deliberate, planned and well thought out methodology that analyzes needs. It subsequently formulates the required elements of design, starts the transformation process by developing the intended thought out design and goes through a deliberate implementation process that should include a formative evaluation process and pilot testing. It then finally arrives at the intended product having completed the summative processes. If one cuts this process short, one is asking to fail. This is why learning professionals involved in providing materials for MET must be certified.

Considering the concept of a global cloud computing system with information and knowledge management as its core, synergies could take advantage of the cloud wherein learned individuals contribute and collaborate for self-benefit and society as a whole. This does not seem possible in a traditional pedagogical framework. Real learning, connectivism, innovation and leap ahead development is what adult learning theories require in this new cxMOOC environment.

These VLE and Cloud Computing Systems (CCS) can reach out and engage global institutions, staff, students and stakeholders who interact on the network as nodes to collaborate, leveraging 21st Century Technology Skills and emerging technologies to do so. This would also help to create momentum towards achieving organizational metrics captured in Balanced and Operational Scorecards, Strategic Initiatives, Key Performance Indicators and the like [11]. Technology best practices can also help articulate and align the institution's goals, vision and mission giving concrete steps and methods with which to facilitate metrics, feedback, realignment and achievement of desired end-states.

5. MOOCs IMPLEMENTATION IN MET

There are any number of regimes, models and schemes that attempt to describe how to use e-learning, i.e., traditional pedagogy online. A vanguard number of talent development professionals (TDP) have moved on from pedagogy outright and have begun to formulate, if not already, robust theories and practices involving andragogy, adult learning, experiential learning and a host of other adult learning regimes suitable in today's global environment [4]. TDPs have also adopted a number of e-learning technologies and platforms in an effort to model these updated traditional learning methodologies online. But, according to McKenna [10], not all improvements in technology equate to significant advances in education.

In fact, in the U.K., to help ensure the correct online pedagogical frameworks or models, it's mandatory for the individuals completing their certificates, diploma, or higher degree to have what is called a Personal Development Plan (PDP) e-Learning Portfolio. This PDP is considered to help enable students to plan, reflect and record progress of their learning experience online while working on their assessment coursework tasks [17].

When taken together in context of the previous discussions and referring to the Advanced Cloud Computing Model (AC2M) in Figure 6, these PDP e-Learning Tools might be an assessment enabler and model that the maritime higher authority can use to approve online platforms. There is documentation that learning occurs.

According to ATD, the Rummler and Brache Nine Performance Variables Model shown in Figure 7 below, considers the three levels levels of performance, which are organization, process and job performance. After which, a Human Performance Improvement Model such as the ASTD HPI Model is appropriate in Tier 1 of Figure 6 to determine the performance gap, root causes and proposed solutions [6].

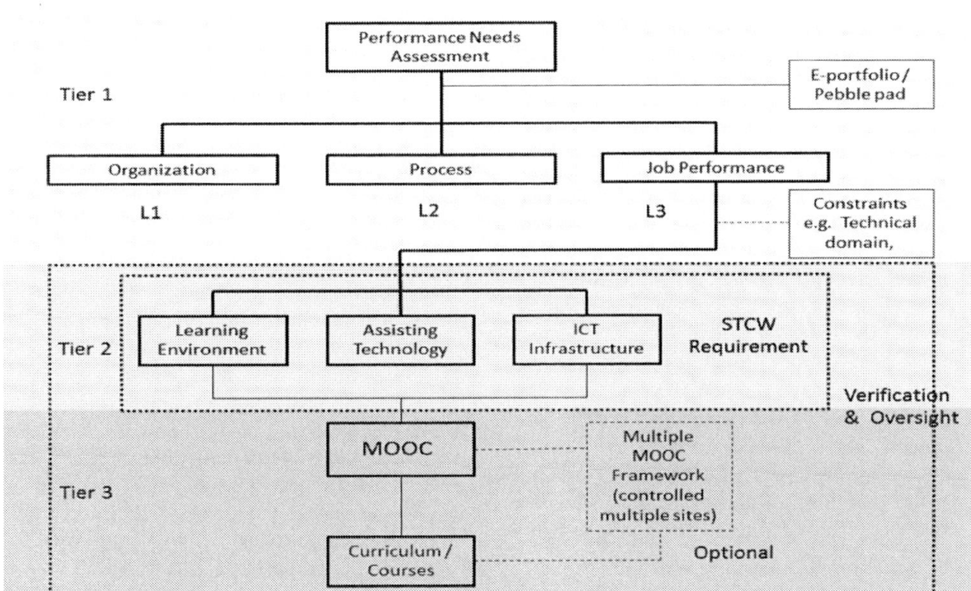

Figure 6: AC2M Model.

THE THREE PERFORMANCE NEEDS

THE THREE LEVELS OF PERFORMANCE	Goals	Design	Management
Organization Level	ORGANIZATIONAL GOALS	ORGANIZATION DESIGN	ORGANIZATION MANAGEMENT
Process Level	PROCESS GOALS	PROCESS DESIGN	PROCESS MANAGEMENT
Job/Performer Level	JOB GOALS	JOB DESIGN	JOB MANAGEMENT

Figure 7: Rummler and Brache Nine Performance Variables [6].

As to forms of oversight, there are LMS systems (e.g., WordPress) that can facilitate multiple sites whose structure can help standardize and set structured websites/LMS (institution sites), ensuring standard LMS themes, templates, curriculums and approved content for development, implementation and oversight. Such a system could also help facilitate assessment, evaluation, auditing and oversight. A central WordPress site would have and can see all institutions as they were created and centrally administrated from that site. However, administering 100 sites is asking a lot and, therefore, after setting up the site, it is recommended to allow certain controls locally for daily administration. It can still be monitored and have the capability to control all functions. This kind of system could work, but is a bit controlling for institutions to buy into the MOOC and LMS schema, but might be necessary if maritime authorities are expected to buy into MOOCs; the price of doing business online.

Regional or global maritime authority authorized testing centers or sites and assessment cooperatives or centers would also be a part of this schema. In Figure 6, one can see verification and oversight in the green shaded area next to the optional MOOC multi-site box. This implies a regime that gives oversight that might involve allowing time to come into compliance with such a scheme, give visits, help and grooming inspections, etc. Other LMS systems such as Moodle and Blackboard are still relevant and may reside in a different layer with the MOOC multi-site on top; links under the WordPress system. Like Moodle, WordPress is also Open Source and can be free, but most likely it would be managed by a third party unless the skills resided locally. The author had set up such a site using WordPress along with a WordPress blog to compliment and encourage collaboration, learning and discussion (see https://globalmetblog.imanfiqrie.com). The WordPress LMS theme is a paid theme. Learning tools like the PebblePAD (an e-Portfolio), helps students to plan, reflect and record [17] their progress in the MOOC. It could be used sort of like the Training and Assessment Record Book (TARB) concept used in the maritime industry to document and facilitate learning (e.g., experiential learning). Authorities could also use this as evidence that learning and assessments (formative) are occurring.

6. BARRIERS IN IMPLEMENTATION OF MOOCs IN MET

Every system has barriers. This might be even more so when it comes to developing and implementing MOOCs in MET. Many stakeholders appear not to be keeping up with technology and this is probably one of the biggest barriers to implementing MOOCs and new technology in MET.

Depending on how one looks at it, one other significant barrier in the maritime industry is MET itself as a whole. Many learning systems and theories in use might be considered antiquated by many standards which includes the pedagogy, poor execution of learning theories, plagiarism, copyright violations and methods as examples. One might also argue that this is a contributory factor to the more than 100 major shipping casualties per annum, as shown in Figure 8 refers [18]. The elements of the proposed maritime learning system should contribute to the stated performance goals of the IMO which are a safe, secure, energy efficient shipping on cleaner oceans [1]. When designing a learning system, as highlighted by ASTD, context is everything [6].

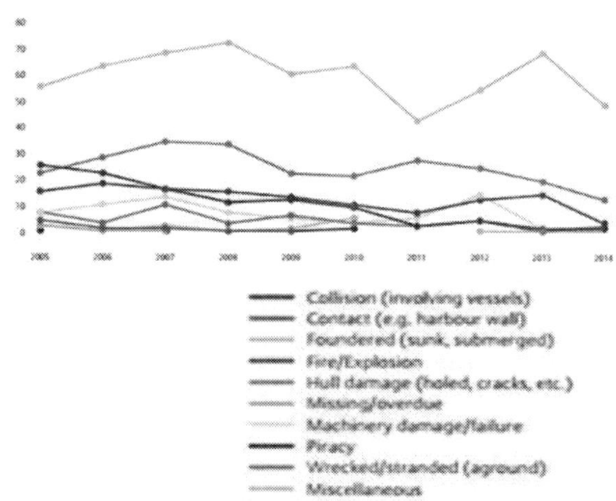

Figure 8: Causes of Total [shipping] Losses 2005-2014 [18].

Other MOOC barriers would include addressing change management and mind-set behavior in both MET and the maritime industry. Global, regional and local cultural differences amount to huge barriers and differences in the development and implementation in MET. Another is that the IT infrastructure capabilities and differences make standardizing MOOC and LMS frameworks in the maritime industry difficult at best and unmistakably is one of the biggest hurdles affecting most, if not all of MET and industry outcomes which includes maritime authority involvement and approval of online curriculums for use in the STCW system.

Throughout all three tiers of the AC2M Model shown in Figure 6, technology is the common thread that binds them. This implies a curriculum that includes the development of the required digital literacy and skills to facilitate the requirements of the study. Many Western entities have already adopted 21st Century Skills curriculums as a partnership involving among other things IT, digital literacy and internet access [19]. It is about time for MET to do something similar. From U.K. to Shanghai, Toronto and even Malaysia and beyond, governments recognize the need for 21st Century Skills [20]. The time has come for the global community at large to get onboard.

There is also something called "technical domain based services" that would be beyond the control of MET and facilitators that amount to significant barriers for consideration. These include (1) IT support; (2) Access (on and off site); (3) Virtual Learning Environments (VLE), third party software and enhancements, some free and some not; (4) software and expertise availability and training; and (5) licensing [10].

Many lecturers can appreciate and have first hand knowledge of IT support issues. However, it is important to try and cooperate with ICT to help ensure they receive training as well and make them realize that they are part of both the discussion and solution. Just the same, MET should not expect ICT personnel to articulate MET's case for them, know for example, all about cMOOCs, xMOOCS, create STCW aligned content and upload that content into a MOOC LMS. The IT piece is a major tool or learning technology and is a necessary and sufficient condition for the success of development and implementation of MOOCs in MET. ICT should primarily ensure infrastructure, equipment, bandwidth, access (on and off-site), WIFI, facilities, software acquisition and usage, VLE and licensing are conducive to the high performance needs of MOOCs in MET. The ICT may not have all the necessary vested interest, assets, knowledge, skills or commitment to also do the learning professional's portion of the development and implementation of MOOCs for it is not their job, but working together will ensure synergy. It is the Training Development Professional's (TDP) job to ensure this happens with ICT support.

MOOCs must be accessible or there is little point in having them. Access literally means anywhere and anytime and if that can not happen, then something is wrong and needs to be corrected as soon as possible. This means between ICT and the TDP, the requisite subject matter expertise must be available. For example, access problems because one is onboard ship is a barrier and must be addressed. This may also mean some expertise must reside onboard ship. The issues caused by time and place present potential licensing issues with those at sea versus not at sea or off-site as participants may not be registered at the main site facility and thus the use of licensed bands of seats becomes a factor when using a paid LMS like Blackboard (Blackboard Moodlerooms possibly the exception) or Adobe Prime LMS.

What VLE (LMS/MOOC) and third party software (e.g., Adobe Creative Cloud) to purchase has been the central question for many institutions for many years; many checklists and weighting schemes exists. This goes back the the level of knowledge and required certification of the TDP. Truth be told, there are more than enough systems that would facilitate the quality process of development and implementation of MOOCs in MET. For example, in the author's opinion, Moodle or Blackboard or even WordPress fits the criterion quite nicely. These are very high-end products even though MOODLE can be Open Source or free; most likely it will be managed (hosted by a company and cost money). One literally can not go wrong with either of the aforementioned choices. It is a matter of the other parameters like certified TDPs, change management, oversight, infrastructure, etc., that will determine success, not necessarily the LMS of the three just mentioned.

Then there is the matter of content development software and subject matter expertise (SME) to use it in the proposed environment. This paper suggests software for the development and implementation of content aside from that possibly available in the LMS. They are not the same, but should be complimentary. For example, Moodle has its own content development system that is outstanding for the inclusion and tracking of objectives, grades, assessments, rubrics and more, but it is not the same as developing flash content, podcasting and other media learning technologies to help transform the learning into required outcomes. These tools would include such things as the Adobe Creative Cloud series encompassing everything from the full Adobe Acrobat to Photoshop and Dreamweaver. In this regard, Adobe Creative Cloud and Captivate (separate products) stand alone as tools the TDP would want to have, Adobe Captivate integrates with Powerpoint; so one opens PPT with Captivate features. These products are high-end and require intermediate to advanced level training and a lot of reading to become proficient.

So far the discussion has been mostly about Tier 2, Complex System Forces and the technology aspect like learning environment, assisting technology, ICT infrastructure and STCW requirements. As discussed, there needs to be some regulatory oversight as well. As can be seen in Figure 6, this section is shaded red in the AC2M for a reason as it can be a show stopper for the MOOC.

There is a saying that 'a good plan executed now is better than a perfect plan executed late'. If it is late, its time has most likely already passed and it is not needed anymore. The plan becomes obsolete if left to chance as are a number of the MET plans regarding emerging technologies already in place today and in need of overhaul.

While a number of models and theories have been presented, the UK Marine Coastguard Agency (MCA) Marine Guidance Note (MGN) 092 could also be considered as supplementary reference. It speaks to the hard details of courses, sea time and maritime authorities. This must be considered when building curriculums, seeking Cadet sponsorship and possible ship berths for the seatime requirements.

7. CONCLUSION

One can conclude that Systems Theory, to include Chaos and Complexity Theory, have a hand in the development and implementation of MOOCs in MET. SMART (specific, measurable, attainable, realistic, time based) goals and solutions have been presented and laid out in this paper. Oversight will be required and is the price of doing business and operating in the modern complex world. There are of course many moving parts proposed here and 'tweaking' any one part may profoundly affect another part of the system one had not anticipated or intended. This is Systems Theory, which also tells us that even in chaos there is order in its complexity, a rhythm. Some cyclical parameters when taken as a whole into account facilitates the faired line brought about by planning and needed outcomes. That faired line must emanate from a thorough and proper assessment, not firstly a training needs assessment. The purpose is to ascertain whether or not an organizational, process or job performance level gap exists. Premature solutions that accurately describe a training need can also be precisely wrong. The priorities are called into question. Plans and solutions must be measured twice and done right the first time using the tools of the learning professional, like ADDIE and other learning professional competencies such as the ASTD Competency Model profited.

The initial assumption that an LMS and MOOC would help seafarers to meet their goals is a bold Systems Theory variable as it has never been done before and is not a certainty, only highly likely. Professional oversight is required to increase the likelihood MOOCs will succeed. A modified AC2M has been put forth (Figure 6) to give a framework and some order in a complex system to a system that includes a multitude of teaching methodologies, theories, technologies, regimes, regulations, authorities and participants. This reinforces the right minded notion that trainers must have more than just a working knowledge of all the aforementioned, in whatever form oversight determines that to be. Just knowing about the STCW in terms of training and assessments required is not enough and a more comprehensive and professional certification encompassing the organization, its processes and the job performance levels is necessary.

These solutions and processes are again undoubtedly complex in nature, making the easiest and shortest path to resistance a likely one. Maintaining a relatively open maritime system of individuals and authorities willing to change as described, is the path of leadership and vision for the maritime future. The missions, goals and priorities between industry, authorities and IMO must properly align to the reasoned approach. IT systems in Tier 2 (Figure 6) should be able to cope with the advanced technologies of the present that some say are coming at a pace likened to Moore's Law. Trainers are required to be skillful in the mechanics of Tier 1, 2 and 3 (Figure 6) to include learning technologies involving the administration and hosting of LMS and MOOCs. Finally, an effective verification and oversight of executive stakeholders at the organizational level across the whole systems towards all the levels of the Rummler-Brache and Tier process as described should be essential.

8. REFERENCES

1. International Maritime Organization. http://www.imo.org/. Date accessed: 29 July 2016.

2. JOHN MAK, S.F., WILLIAMS, R. & MACKNESS, J., 'Blogs and Forums as Communication and Learning Tools in a MOOC', *Proceedings of the 7th International Conference on Networked Learning*, 2010.

3. KOP, R., 'The Challenges to Connectivist Learning on Open Online Networks: Learning Experiences During a Massive Open Online Course', *International Review of Research in Open and Distance Learning, Vol.12, No.3, 2011*. Date accessed: 15 July 2016.

4. FIQRIE, I. & YUTUC, W., 'Use of Virtual Learning Environments and CLoud Computing Systems in Maritime Education and Training: Issues and Challenges', *Education & Professional Development of Engineers in the Maritime Industry. The Royal Institute of Naval Architects. April 2014. (pp. 109-116).*

5. FIQRIE, I., 'Cafe Iman West M31 LMS', *http://moodle31.imanfiqrie.com*. Date accessed: 7 July 2016.

6. ASTD. The ATD Learning System: The Official Resource for CPLP Study. *ASTD Press*. 2013.

7. TRACHTENBERG, S. & PIERCE, S. R., 'Governance Reconsidered: How Boards, Presidents, Administrators, and Faculty Can Help Their Colleges Thrive', *Jossey-Bass*. 2014.

8. FIQRIE, I., 'Fedena School Management System', *http://imanfiqrie.com:3000* . Date accessed: 9 July 2016.

9. MAZOUE, J. G., 'The MOOC Model: Challenging Traditional Education', *EDUCAUSE REVIEW*, N.p., 28 Jan. 2013.

http://er.educause.edu/articles/2013/1/the-mooc-model-challenging-traditional-education . Date accessed: 3 August 2016.

10. McKENNA, G. F., & BAXTER, G. J., 'A Systematic Map of Evaluation Criteria Applicable for Evaluating E-Portfolio Systems', *In V. C. X. Wang (Ed.) Advanced Research in Adult Learning and Professional*, 2014.

11. P21 PARTNERSHIP FOR 21ST CENTURY LEARNING, 'P21 Common Core Toolkit', 2011. http://www.p21.org/storage/documents/P21CommonCoreToolkit.pdf . Date accessed: 2 August 2016.

12. MALAYSIAN EDUCATION BLUEPRINT 2013-2025, 'Preliminary Report Malaysian Education Blueprint 2013-2025', *Ministry of Education, Sept. 2012*. http://www.moe.gov.my/userfiles/file/PPP/Preliminary-Blueprint-Eng.pdf . Date accessed: 1 August 2016.

13. IMO, 'The STCW Convention and Code 2010 Manila Amendments', http://www.imo.org/en/OurWork/HumanElement/TrainingCertification/Pages/STCW-Convention.aspx , Date accessed: 29 July 2016.

14. RAUNEK, 'Top 5 Online Maritime Courses Providers', *Marine Insight*, http://www.marineinsight.com/careers-2/top-5-online-maritime-courses-providers/. Date accessed: 6 July 2016.

15. IMO, 'International Shipping Facts and Figures – Information Resources on Trade, Safety, Security, Environment', *Maritime Knowledge Centre, p17*, 2012.

16. BATES, A.W., 'Teaching in a Digital Age', *BC Open Textbook*, https://opentextbc.ca/teachinginadigitalage/ . Date accessed: 14 July 2016.

17. ORSINI-JONES M., 'Latest Case Studies', *Integrating PDP (Personal Development Planning) and the the E-portfolio PebblePAD*, 2008. http://www.english.heacademy.ac.uk/explore/publications/casestudies/technology/eportfolio.php . Date accessed: 2 August 2016.

18. SAFETY AND SHIPPING REVIEW, 'Munich: Allianz Global Corporate & Specialty', Mar. 2015. PDF.

19. P21 PARTNERSHIP FOR 21ST CENTURY LEARNING, 'Overview of State Leadership Initiative - P21', http://www.p21.org/members-states/partner-states . Date accessed: 2 August 2016.

20. STEWART, V., 'Assessing 21st-Century Skills and Competencies Around the World', *Asia Society*, (http://asiasociety.org/global-cities-education-network/assessing-21st-century-skills-and-competencies-around-world) Date accessed: 20 Aug. 2016.

9. AUTHORS' BIOGRAPHY

Iman Fiqrie B Muhammad@William E. Hamilton, A.A., B.S., M.B.A., holds the current position of lecturer in the Modular Offshore Safety Department at Malaysian Maritime Academy (ALAM). He is responsible for the conduct of lectures and practicals for the STCW Basic Training course, Proficiency in Survival Craft and several other offshore training courses. He is a retired U.S. Navy, Lieutenant Commander with more than 12 years of sea duty and more than 23 years of active service - to include nuclear submarines and combatants. His previous experience includes the hosting of several websites - including the VLE Moodle, content management system Drupal 7, WordPress, others and actively participates in campus e-learning initiatives; Google Apps for Education Administrator, writing articles on e-learning and related matters in the GlobalMET Newsletter, as well as experience working with mobile devices in various settings for the purpose of e-learning. His interests include understanding, creating and implementing successful virtual learning environments, cloud computing and content management systems designed to help facilitate massive open online courses (MOOC) and various other content. Also pursuing a professional certification in learning performance (CPLP).

Wilfredo Erenio Yutuc, PhD (MarEd) holds the current position of Senior Lecturer in the Advanced Marine Engineering Department at Malaysian Maritime Academy (ALAM). His responsibility is to conduct lecture and assessment to marine engineers who are preparing for their Class 2 and 1 Certificate of Competency (Motor and Steam) on unlimited voyage of 3000 kW or more. He has more than 20 years of experience in marine engineering, 10 years of which was spent onboard as an engineer and more than 10 years on various positions in ship management and maritime education and training. His research interest is on green ship technology, shipboard energy efficiency and use of technology in education.

SKILLS AND ASSESSMENT NECESSARY IN A CO-OPERATIVE EDUCATION MARITIME ENGINEERING PROGRAM

M Symes, Australian Maritime College, University of Tasmania, Australia

SUMMARY

Maritime industry and academia constantly battle to provide seamless transition from undergraduate to professional engineer. Co-operative education (co-op) programs endeavour to bridge this divide by ensuring undergraduates experience current industry practices and procedures throughout their undergraduate program, confirming co-op graduates ability to 'hit the ground running' upon completion of undergraduate studies.

Success of co-op type programs as a form of Work Integrated Learning (WIL) is reliant on structured and authentic assessment to ensure development of required skills. The challenge exists to ensure this is maintained within work placements which occur outside the academic institute. This paper investigates the processes utilised within the National Centre for Maritime Engineering (NCMEH) at the Australian Maritime College (AMC).

1. INTRODUCTION

The Department of Maritime Engineering at the Australian Maritime College currently provides a co-operative education program to students enrolled in four-year full time Bachelor of Engineering Degrees in Naval Architecture, Ocean Engineering and Marine and Offshore Engineering. The program is not compulsory and is only available for high performing students.

The program is entering its 4th year of operation and has undergone significant modifications during its short lifespan. Modifications both in terms of assessment methods and work placement structure have been necessary based on feedback from industry and to ensure program continuation within a department with limited resource.

The program is unique in its ability to provide co-op students with four workplace opportunities within the standard undergraduate program. Although the academic program is unchanged some changes have been made to the delivery timetable which can be seen in Table 1 below has necessitated an increase in the prerequisite needs of students that choose to enrol in the co-op program compared to the standard degree program.

Table 1: Co-operative Education program structure

	Nov-Feb	Feb-Jun	Jul-Nov
Yr 1	Current 1st Year Academic Program		
Yr 2	Work Term 1 (TasTafe)	Academic Semester 2/1	Academic Semester 2/2
Yr 3	Work Term 2		Academic Semester 3/2
Yr 4	Work Term 3	Academic Semester 3/1	Work Term 4
Yr 5	Work Term 4 – cont.	Academic Semester 4/1	Academic Semester 4/2

Australian academic institutes operate on a two semester academic program which provides another challenge for students in a global employment environment. An example is AMC students applying for North American or Canadian placements which operate on a different (three semester) academic structure.

Due to challenges faced placing first year students within industry after only one academic year, work term one has been modified to be a competency based assessment of engineering practicum module offered by Tasmania Tafe (TasTafe). The TasTafe modules include general machining, thermal cutting and manual metal arc welding as shown in Table 2 below. This engineering practicum unit allows students to explore aspects of engineering without the specific focus purely on maritime engineering.

Table 2: Co-operative Education program structure

Module	Descriptor
MEM05004C	Perform routine oxy acetylene welding
MEM05007C	Perform manual heating and thermal cutting
MEM05012C	Perform routine manual metal arc welding
MEM05051A	Select welding processes
MEM05052A	Apply safe welding practices
MEM07005C	Perform general machining
MEM12023A	Perform engineering measurements
MEM18001C	Use hand tools
MEM18002B	Use power tools/hand held operations

During the first academic year of the program co-op students are also introduced to a professional development program to assist with Curriculum Vitea (CV) development, interview skills and an introduction to workplace expectations.

This first year structure provides multiple benefits to the overall program. Firstly it ensures that students are given time to adjust to academic life both in terms of personal changes but also in terms of understanding the requirements of academia. Secondly it allows academic staff to monitor the academic progress of all students to ensure performance requirements are maintained. This allows staff to make changes to course enrolments up to the completion of two academic years with minimal effect on student progress as TasTafe units are planned to be credited as a student elective if course transfer is required.

Challenges exist in ensuring academia can assess a unit which is run totally outside of the academic institute. In terms of pedagogy research into co-operative education proponents argue that reflection is the strongest learning tool that can be utilised [1]. Research has been carried out in the area of reflection and its use to enhance learning in general [2]. The challenge that exists for NCMEH is that as the program develops is good reflective practice an indicator of improved skill level [3].

2. PURPOSE OF ASSESSMENT

Engineers Australia as the certifying authority for engineering bachelor's degrees within Australia and requires students to achieve a level of competency prior to graduation. Academic institutes endeavour to provide evidence of these levels of achievement via assessments.

By utilising both formative and summative assessment methods, evidence can be obtained to measure the achievement of these competencies at (beginner, intermediate and advanced) levels of competency.

2.1 FORMATIVE ASSESSMENT

Formative assessment is used within cooperative education as a method to assist students in understanding their strengths and weaknesses and allowing students to take control of areas that require focus [4]. This focus can relate to not only the academic program but also inspire students into specific career focus paths.

One method of formative assessment is the use of student journals which allows academic staff to observe development of skills and knowledge against a known timeline provided by date stamped student entries whilst on work placement. The use of journals, reports and portfolios provides evidence of development of competencies at various levels for both the student and academic staff.

2.2 SUMMATIVE ASSESSMENT

Summative assessment allows comparative analysis of student's achievements against a benchmark or standard to determine a point of completion of a course or program [5]. Work placements do not have a required standard placed against them however tasks achieved during work placements can be mapped to competencies expected to be achieved within the academic program.

Technical reports are one such method that can provide an excellent means of assessing students' ability to combine both knowledge obtained within the workplace and that developed within the academic institute.

3. ASSESSSMENT

In an effort to reduce employer workload a final workplace review is the only form of workplace assessment that is carried out. Although this is a summative form of assessment the cumulative result over multiple work placements is deemed formative.

The current assessment structure of the co-operative program can be seen in Table 3 below. All pieces of assessment from all co-op students are submitted within two weeks of return to academic semester. This ensures that all work term requirements have been met prior to academic semester census date.

Table 3: Assessment structure of Co-operative Education program

Work Term	Assessment
One	Practical competency based assessment carried out by TasTafe of engineering based modules
Two	Engineering Competency workbook, Technical Report, Employer Workplace Assessment
Three	Engineering Competency workbook, Technical Journal, Employer Workplace Assessment
Four	Engineering Competency workbook, Oral Presentation, Employer Workplace Assessment

3.1 FACULTY INPUT

Faculty assessment consists of all technical and reflective activities undertaken by students. These activities are carried out independently by the student whilst in the work domain. Students have minimal guidance from academic staff members whilst on work placement. Students are however free to discuss with academic staff during placement the particulars of any technical report if concerns are raised by industry over security or intellectual property ownership.

All reflective material is completed within the students' engineer's notebook which is utilised in both the standard and co-op programs. This provides co-op program coordinators the opportunity in the future to make comparative analysis between standard and cooperative education programs.

Academic staff members make one visit to each student per work term. Priority is given to visiting students undertaking the two longer work placements. Students undertaking international placements currently do not receive visits from academic staff.

3.2 INDUSTRY INPUT

Industry representatives are asked to assess all students upon completion of work placements. To reduce load on industry representatives the review process consists of eight attribute areas that are rated from unacceptable to outstanding. Attributes covered include areas such as accuracy of work, ability to work with others, adaptability, punctuality, adherence to rules and policies. Employers have the option to provide further written comments to validate results. This employer evaluation is backed up by a site visit by co-op program academic staff to ensure alignment of requirements by all parties namely academia, employer/and student.

Industry is asked to involve students in the work placement review of performance process to allow students opportunity to reflect. Industry is also given the option to provide private feedback / comments / recommendations to the co-operative program coordinators relevant to academic curriculum and program structure.

Industry input in this area ensures the academic institute is maintaining a program relevant to the needs of a rapidly changing employment landscape whilst also highlighting to new employers the capabilities of NCMEH students.

Levels of industry engagement do vary based on workload and perceived strength of industry segment at the time. Challenges continue to exist in ensuring industry comprehend the effects on student understanding of cooperative education when industry focus continually varies from educational to economic needs based on market forces [6].

4. MAINTAINING THE BRAND

To highlight the importance of the cooperative education program to NCMEH, students that undertake work placements have the results of that placement recorded on their testamur upon completion of the full program. This provides a method for prospective employers to not only have evidence of academic performance but also evidence of student performance in the workplace. This workplace evidence can also be across multiple facets of the discipline. An example of this may be a student that is placed within design, production and support on three different work placements with three different employers.

This type of branding also increases the exposure of the AMC program throughout different engineering branches without the need for specialist staff or academic staff input. This method is also intended to highlight to all employers the benefits of cooperative education as a selection tool.

Structured placements throughout the degree program also provide evidence to certifying bodies such as Engineers Australia that multiple opportunities are provided for students to improve the level of competency achieved. It is envisaged that reflective type assessments will allow future comparisons between standard and cooperative education type programs.

The methods stated above have been put in place to ensure evidence of separation between the standard engineering degree program and the co-op program both in terms of student recruitment and program certification.

5. MAINTAINING A SUSTAINABLE ASSESSMENT

The challenge for any small institute such as AMC is ensuring continued development of the program with limited support staff. NCMEH does not have special focus staff to support the development of the program. Only academic staff that strongly believe in the benefits of such a learning process have taken on this role. For the program to be successful it is necessary to ensure that this support level can be maintained by the same small group of academics irrelevant of growth in student numbers.

Although work placements take place outside of the academic institute it is important to provide quick turnaround of work placement results to ensure smooth progression of student development. This exists as a continuing challenge as student numbers continue to grow.

Another challenge for NCMEH is engaging students upon return to academia after long work placements of six to seven months within an industry environment. Students regularly find it difficult to readjust to university life after such an immersed experience within the workplace. This is very evident for students completing their final work term where they may be in the situation that their final employer was impressed and may have made an offer of employment although the student is still required to complete a final year within academia. This situation can modify the student belief that learning is still a requirement of their final year within academia.

6. CONCLUSIONS

To ensure survival and continued development of cooperative education it is important to consider employers as partners in education curricula and not just recipients of outcomes [7]. This partnership if communicated well will result in graduate engineers with skills relevant to industry at the time of graduation.

Cooperative education within maritime engineering in Australia is in its infancy and requires the support of larger multinational companies that have had exposure to other forms of work integrated learning. Continued exposure to co-op students will provide industry with evidence of student capabilities and skills at various stages of their degree program. Strong industry participation will ensure students feel a sense of inclusion and that their input is positive and meaningful leading to confidence and a desire to continue along a path of lifelong learning [8].

7. REFERENCES

1. COLL, R. K., EAMES, C., PAKU, L., LAY, M., HODGES, D., BHAT, R., RAM, S., AYLING, D., FLEMING, J., FERKINS, L., WIERSMA, C., & MARTIN, A. 'An exploration of the pedagogies employed to integrate knowledge in work-integrated learning', *Journal of Cooperative Education and Internships,* 43(1), (pp 14-35). Wellington, New Zealand, 2009.

2. BOUD, D., 'Sustainable assessment: Rethinking assessment for the learning society' *Studies in Continuing Education*, 22(2), (pp 151-167). London, UK, 2000.

3. RAELIN, J. A., GLICK, L. J., MCLAUGHLIN, K. E., PORTER, R. D., & STELLAR, J. R., 'Reflection-in-action on co-op: The next learning breakthrough', *Journal of Cooperative Education and Internships.* 44(2) (pp 9-15). Myrtle Beach, SC, USA, 2008.

4. HODGES, D., 'The assessment of student learning in cooperative and work-integrated education.' *International handbook for cooperative and work-integrated education: International perspectives of theory, research and practice* (2nd ed. pp 53-62). Lowell, MA, USA, 2011.

5. BOUD, D., & FALCHIKOV, N., "Aligning assessment with long-term learning." *Assessment & Evaluation in Higher Education* 31, no. 4, (pp 399-413). Oxfordshire, UK, 2006.

6. TODD, A. M., LAY, M., 'Cooperative and Work-Integrated Education in Engineering' *International handbook for cooperative and work-integrated education: International perspectives of theory, research and practice*: (2nd ed. pp 111-121). Lowell, MA, USA, 2011.

7. BRAUNSTEIN, L. A., TAKEI, H, WANG, F, & LOKEN, M. K., 'Benefits of cooperative and work-integrated education for employers' *International handbook for cooperative and work-integrated education: International perspectives of theory, research and practice*: (2nd ed. pp 277-286). Lowell, MA, USA, 2011.

8. HARTE, D., & SYMES, M. 'Challenges of developing and sustaining a Co-operative Education Program in Maritime Engineering' *Proceedings of the AAEE 2013 Conference. 1E (3),* Gold Coast, Australia, 2013.

8. AUTHORS' BIOGRAPHIES

Mark Symes received a Bachelor of Engineering degree in Naval Architecture from the Australian Maritime College (AMC). Mark is a lecturer and Course co-ordinator for the cooperative education program within the National Centre of Marine Engineering and Hydrodynamics (with a specialization/concentration in Engineering Design and Production). His research interest includes Engineering Education with a particular focus on project/problem-based learning. Mark's research highlights the importance of aligning learning outcomes and assessment strategies with authentic real world projects that cater for a wide range of learning styles that promote deep learning. More specifically, his work examines the obtainment and assessment of Engineers Australia's stage one graduate competencies.

MARITIME TRAINING SHIP: A NEW PERSPECTIVE OF STANDARD MARITIME TRAINING FOR A GLOBALLY COMPETITIVE SEAFARER

R H Brigoli, J Antolihao, R Saing, L A Cañete, and J Auxilio, University of Cebu, Philippines

SUMMARY

This study aims to standardize the Filipino seafarers in consonance to the adoption of the comprehensive amendments to the 1978 International Convention on Training, Certification and Watchkeeping for seafarers – know to us all more conveniently as the STCW Convention. To meet the global competence in the shipping industry and in order to produce well-trained seafarers to ensure safety of life at sea, maritime security, efficiency of navigation, expand the volume of Filipino ship officers and reduce the rating Filipino seafarers.

This study will integrate the maritime training facilities on shipboard that will realise the actual scenario of the ship during the span of the training. The ship will serve as a "floating maritime training centre" that will be equipped with facilities to meet the STCW Convention requirements. The ship can also navigate within the Philippine waters and dock in every port of the Philippines to ease the transportation means of the seafarer or apprentice. This study will help realise to meet the deadline of STCW implementation as mandated by the International Maritime Organization and most of all to help in becoming globally and morally competent seafarers in the shipping industry.

NOMENCLATURE

BIMCO	Baltic and International Maritime Council
IMO	International Maritime Organization
STCW	Standardisation of Training, Certification and Watchkeeping

1. INTRODUCTION

1.1 MANNING CAPITAL

Philippines is recognize as a world leader in the supply of personal services. As the manning capital of the world, Filipino seafarers are deployed in almost every vessel sails on the seven seas. These Filipinos are employed because of their capabilities to remain competitive, given the increasing demands of the global market. A proof of acceptability and recognition of the country's educational standards is the increasing number of foreign students enrolled in tertiary institutions. This scenario presents a challenge to our educational system: to gain more international respect and recognition, one should aim for higher international comparability and standards. The challenge of this international standardisation has to be taken seriously if the country is determined to be an active implementer of STCW.

Despite the gap between the financial resources of Philippine universities and those abroad, it is inevitable to do quality assurance as well as international standardization. If the time and resources poured into the educational programs of maritime are effective, the country will have a strong human capital that is able to bring about the desired level of social and economic development essential to strengthen the STCW implementation. What is needed is a thorough study of the design and implementation of maritime programs and accessibility of experience to pinpoint accurately the advantages of having a "floating maritime training centre". The experiences, both positive and negative, of other countries and other maritime educational systems will be of great help to inform policy making and scholarly research.

As higher training standards are mandated, compliance to the measures is obligatory to all signatories to the convention, of whom the Philippine is one. It was the aim of the study to standardize the Filipino seafarers in consonance to the adoption of the comprehensive amendments to the 1978 International Convention on Training, Certification and Watchkeeping for seafarers. Thus, the study will be an avenue to local maritime training centres and regulatory agencies to achieve an uncompromising attitude toward quality upgrading.

2. CHALLENGES EXIST FOR THE PHILIPPINE CREWING INDUSTRY

BIMCO recently released their 2015 Manpower Report on the global demand and supply of seafarers. There are 1,647,500 seafarers inclusive of both officers and ratings on earth. Out of all these seafarers, The Philippines provides the largest number of ratings. Despite the large number of Filipino ratings, the country only ranks number two—next to China—as the world's largest provider of seafarers across all ranks.

Hartmann Crew Philippines jot down different factors to enhance the movement in ranks of the Filipino ratings.

- **Enhanced Maritime Education** – Better foundations from educational institutions will ensure that future maritime professionals are equipped with the right knowledge.

- **Career Development** – Agencies employing seafarers must take an active part in developing the careers of its crew in order to contribute to their growth as professionals.

- **Training** – Continuous learning is crucial. Technology onboard ships change rapidly and seafarers going up the ranks must learn to keep up. It's also an advantage for seafarers to deepen their knowledge on the changing economics and multi-cultural aspects involved in the shipping industry.

- **Competitive Wages** – Filipinos face competition from other nationalities for a place onboard. With the current downtrend of the economy, ship owners are least likely to pay for higher wages when more competitive rates are present elsewhere.

- **Quality Performance** – There is no substitute for quality. Filipino crew must continue to deliver excellent work and service in order to cement their place in the industry and to justify the cost of employing them onboard ships.

"The reality is that there is a challenge at hand for the Philippine Maritime Industry. A challenge that can turn into an opportunity if the seafarers together with the administration and the rest of the industry players are willing to augment the system and integrate these factors in order for the Philippines to become the first choice of highly qualified maritime professionals regardless of rank", (Hartmann Crew Philippines).

3. CONCLUSION

Training facilities for Seafarers are limited and insufficient in the Philippines, students travel from one place to another to render the training needs. Financial and security are two common reasons why nautical students quit form their trainings and rating seafarers stays as they are. This era of competition should not be taken for granted.

Statistics shown, Philippines was the manning capital for low earner seafarers (Figure 1). A Study tells, by 2025 there are thousands of ratings will be left behind and shortage of officers (Figure 2). This trend can be cease if there is a real execution of the factors cited by team Hartmann Crew Philippines.

Researchers interpolated; A Training Vessel which collects apprentice from designated ports in the Philippines will lessen the worries of those not capable of travelling to the cities with facilities of their own. Passengers on board are continuously trained while accumulating the new ones; trainees will gain more actualisation than on the land training facilities since they are trained on the sea. Thus, STCW will be easily implemented, high chance of upgrading ranks of the ratings, cut-off shortage of officers and no one will be left behind the voyage.

Figure 3, is a glimpse of the floating training facility for the Filipino seafarers.

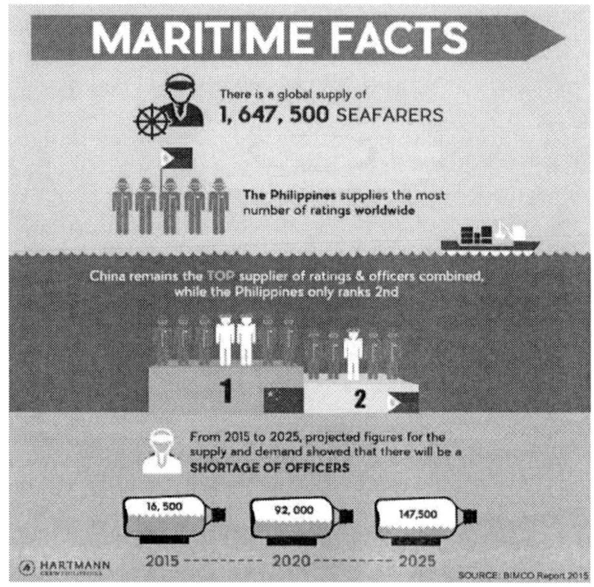

Figure 1: Philippines versus China Supply of Officers and Ratings

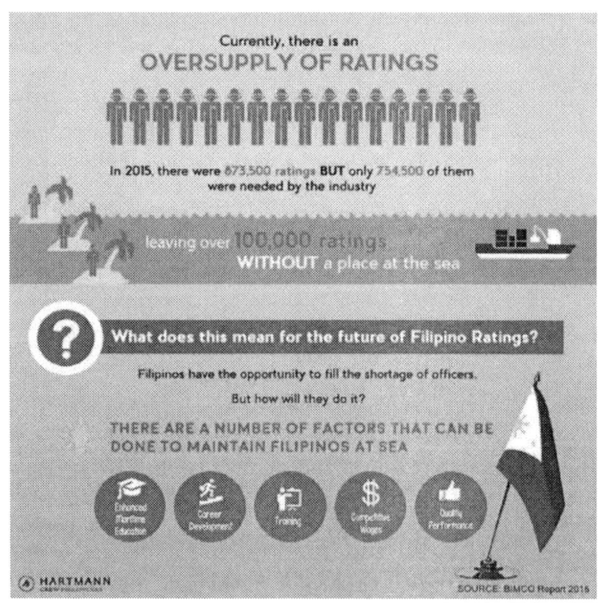

Figure 2: Shortage of Officers in 2025 due to Oversupply of Ratings

Figure 3: Marino-Filipino Training Ship

4. REFERENCES

1. Maritime Facts, Hartmann Crew Philippines,(2015)

2. TUPPER, E. C., Introduction to Naval Architecture (4th Ed.). *Great Britain*,(2004).

3. DR. BARRAS, C. B., Ship Design and Performance for Master and Mates (1st Ed.). *Great Britain*,(2004).

4. TAYLOR, D. A., Introduction to Marine Engineering. (2nd Ed.),(1996).

5. MCGEORGE, H. D., Marine Auxiliary Machinery (7th Ed.), (1995).

6. EYRES, D. J. & BRUCE, G. J., Ship Construction (7th Ed.), (2012).

7. The SNAME., Ship Design and Construction (Vol. 1),(2003).

8. Graphical Symbols for Shipboard Fire Control Plans, *International Maritime Organization*, (Res. A.952), (2004).

9. Rules for Building and Classing Steel Vessels under 90 meters in Length, *American Bureau of Shipping*,(2015).

10. Philippine Merchant Marine Rules and Regulations, *Maritime Industry Authority*, (1997).

11. Philippine Maritime Accidents, *Philippine Coast Guard*,(2016).

12. C/M SABORDO, H. J. & BUGA-AY, Maritime Accidents in the Philippines from 2000 – 2010, A Documentary Analysis, (2013).

13. CHUNG DO NAM, A Study of Shipboard Training for the Efficient Maritime Training. (1995).

5. AUTHORS BIOGRAPHY

Hughes Rodolfo A. Brigoli is a B.S. Naval Architecture and Marine Engineering Student at University of Cebu, Philippines. Principal author of the study during his research design subjects last S.Y. 2015-2016. He is currently affiliated at Lite Shipping Corporation as a Part-time Naval Architect based at Cebu City, Philippines. His experiences earned him a lot of knowledge and wisdom which help him to create this study. He worked previously at Maritime Industry Authority (MARINA) Davao Regional Office, Davao, Philippines and Nagasaka Shipyard, Inc., Consolacion, Cebu, Philippines.

Jacklyn R. Antolihao is the present chairperson of Naval Architecture and Marine Engineering Department in University of Cebu, Philippines and a licensed Naval Architect and Marine Engineer. As a department head, it is her duty to guide the students and faculty as well as moulding the entire team ready for a life-long sustainability. Her Shipbuilding experience for Steel Vessels, Fibre Reinforce Boats, and shipping consultancy were the minister to her advice and collaboration as co-chief author in conception of the study.

Larry Angelo R. Cañete, chairman of Electrical Engineering Department in University of Cebu-Main Campus, Philippines. As an Electrical Engineer, his collaboration to the study focus on the electrical system layout, maintains electrical control systems and components to its required specifications and recommends suitable actions to minimize or eliminate recurrence of problems. Formulate preventive maintenance schedules and associated procedures and checklist related to electrical aspects.

Richard t. Saing, dean of the College of Engineering, University of Cebu-Main Campus, Philippines and a Licensed Electrical Engineer. His experienced in United Nation Industrial Development Organization as Country Manager on Energy Management & Systems Optimization for large industries brighten the concept for vessels propulsion efficiency. Seeing as his research interests were energy efficiency, renewable energy resources, power systems protection and stability, distribution and transmission systems, power systems economics, energy economic dispatch and power quality it is rest assured the training vessel will sail on high seas with great economic value and sustainability.

Jared C. Auxilio, licensed Naval Architect and Engineer. Part-time instructor of Naval Architecture and Marine Engineering Department, extending help in realisation to concrete ideas is one of his responsibilities towards the students. Who is now currently connected to local ship operator, Gothong Southern Lines Inc. As Drydocking Superintendent, he's also a consultant and estimator for related field of expertise. Both his experience in local and abroad help stabilise the design of subject studied.

PROJECT MANAGEMENT TRAINING BY MANAGERS, FOR MANAGERS

Dr M Thurlbeck, Stapleton International Pte Ltd, Singapore

SUMMARY

This paper describes the contextual development of the collaboration agreement recently signed between Stapleton International and the University of Sunderland. The content provides details of the initial identification of the requirement for a different style and content of a Project Management Short Course which was resultant from extensive assignments within the Marine and Oil & Gas sectors. The document outlines the market gap in short course provision which recognises the need for a more sector driven delivery of knowledge transfer, rather than standard offerings of many academic institutions and training providers. It then goes on to describe how the development of a suitable product was achieved whilst conforming to the requirements of QAA and the legal department of the University. Finally, the opportunity to widen the course directory is discussed before concluding with recommendations of proposals for future growth within this area.

1. INTRODUCTION

Stapleton International is a leading edge consultancy providing service solutions to the Engineering, Construction, Marine and Oil & Gas Industries. Since 1972 to the present day they have developed expertise in management, engineering, quantity surveying, planning and legal professions. Now, in collaboration with the University of Sunderland, they have developed a Project Management Training Programme to be delivered in a modular format which is designed with basic and advanced levels. This exciting new initiative will facilitate the transfer of knowledge and experience of Lessons Learned to both individual and corporate students.

Stapleton International have recognised the need for vocational, practical oriented and accredited project management courses. There is also a need to train junior, existing and potential project managers in more effective and advanced leadership of projects in complex multi-cultural and multi stakeholder environments that now exist for many major organisations. There are immeasurable benefits available through learning about sector best practice and also by having the courses accessible to both companies and individuals alike.

2. EXISTING PROVISION

Whilst there are many Project Management Courses available, it is a well-recognised understanding that many industry veterans are retiring and taking with them skills, expertise and wisdom. It takes an average of eight to ten years for a worker to develop that kind of expertise [1]. Hence the question consistently asked was what happens to the future workforce when many veterans are no longer active within the respective industry? How can such knowledge be retained and be passed on to the new generations of those individuals involved within project management? The Society of Petroleum Engineers estimates that up to 50% of skilled workers could retire within the next five to seven years, leaving a large hole in the pool of experienced talent available [2].

A plethora of Project Management courses exists, the majority of which concentrate on the theory and specified tools to achieve successful execution. Many such courses are focussed on standard management theories, philosophies and models. However, as we have all experienced, projects tend to be a little unpredictable as they progress. All too often an unexpected event will create a problem within some aspects of the programme. In many such cases there is often an absence of a definitive theoretical solution and corrective action is only achieved by access to experience. This is well illustrated when considering the quotation "In theory, there is no difference between theory and practice. In practice, there is." [3]

Hence the course was developed based on the experience of Stapleton International and the University of Sunderland enabling an advanced programme which is a combination of theory and practice.

The course is delivered in two parts, basic and advanced and was designed and developed by experienced Project Managers and is delivered by experienced Project Managers. The target audience for the basic module is for those individuals who an organisation believes will progress to project management and leadership roles. Alternatively, it could be individuals who are already in those roles but are in need of training and refreshment and/or assessment of their suitability.

Advanced modules will include those individuals who are already performing well and will be asked to take on bigger and more complex challenges. Key targets are project engineers, project control engineers and managers, QHSE engineers and design engineers. Students completing the course satisfactorily will be awarded an attendance certificate by the University which will also add evidence for individual CPD assessment targets.

However, there were quite a number of hurdles to overcome before the final programme was completed allowing the collaboration agreement to be finalised and agreed. Perhaps the biggest element to be addressed was achieving the sanction of the legal department of the University who among other issues arising in drafting the collaboration agreement, had to ensure the requirements of QAA were satisfied. Whilst this was understandable and there was no intent nor actions taken by Stapleton International to short cut any process, the ability to demonstrate clearly to the Academic Board the need for a novel approach to the short course development proved difficult in selected segments. These are discussed in further detail in Section 6.

3. MARKET DEMAND

Reports from Western Australia predict the shortfall of over 50,000 personnel in the Oil & Gas sector alone if they are to realise the recovery of identified assets [4]. The Middle East have similar shortages, exacerbated by the fact that 50% of skilled engineers are eligible to retire in the next ten years [5]. This is despite a huge reduction of employee numbers which can be directly related to the drop in oil prices over the past two years. Previous reductions in numbers following similar oil price collapses have led to a substantial problem in recruitment of suitably qualified staff when the upturn does occur.

The effect of this skills shortage compounds the ability for knowledge transfer to be satisfactorily completed. In such instances many of the existing management choose to retire or switch industries leading to a shortfall of experienced personnel who can mentor and help nurture existing talent to become the managers and leaders of the future. The rationale for the development of a non-credit bearing short course, initially in Project Management, is as a direct result of a workforce development need identified by Stapleton International within the marketplace. In many of the assignments Stapleton International were involved with, observations were that the majority of projects witnessed suggested they would benefit from some form of structured training to help develop junior engineers and project personnel.

Stapleton International identified a need within large capital engineering and construction organisations who are increasingly looking to train and develop their staff in effective project leadership through external and internal training modules. There is a gap in the market for vocational, practical oriented and accredited project management courses. There is also a need to train junior, existing and potential project managers in more effective advanced leadership of projects in complex multi-cultural and multi stakeholder environments that now exist for many major organisations.

Further research identified the potential for involvement with the provision of skills development in East Africa. Discussions have taken place with leading figures in the respective government departments who have the responsibility for progressing the accreditation and certification of TVET programmes within their respective countries. The opportunities for development of management training programmes in this region far exceeds the original expectations.

By way of a tangible example, The Uganda National Oil Company (UNOC) which only commenced operating on 1st August 2016, recently agreed the Uganda-Tanzania pipeline which is due to start development at the beginning of 2017. It is estimated by the Ministry of Education that up to 30,000 skilled technicians and craft workers will be required for the project. Consequently, the requirement for project management is very high and with little existence of experienced personnel in this area, the shortfall will be made up from ex-pats and locally trained personnel.

4. CURRENT ACADEMIC PROGRAMMES

A search of the web for academic courses identifies that within the UK University system there are over 100 who offer courses in Project Management. These range through short courses, first degrees, post graduate certificates, diplomas and ultimately second degrees. By adding into the equation provision available in USA and Australia, a similar serach identifies there are almost 300 institutions who offer Master's Degree programmes in Project Management [6]. However only five offer a focus on oil & gas. This limit of options was consistently viewed as a failure by the many clients of Stapleton International. Generally, the field of Project Management is traditionally divided into 10 sub-categories [7]. These individual categories form a good basis for standard modules.

The completion of a comprehensive Gap Analysis was the starting point for discussions between Stapleton International and University of Sunderland Programme to review the possibility of developing a different strategy to fulfil the perceived requirement of a new short course capability.

Following several exploratory meetings with various key figures within the University is was agreed that further work should be undertaken to establish (a) if there was indeed a market for a short course being oil & gas specific and (b) what would be the main elements of such a course. Gathering data and comments from clients over a period of around six months, the results confirmed the potential opportunities for business growth following the introduction of a new initiative.

Stapleton International at this time were already delivering workshops and seminars on Contract Risk and Turnaround/Shutdown Management. Senior management considered serious attention should be given to the introduction of a Training Division within the company as this could become a worthwhile business stream providing additional revenue.

The key factor is that the programme has been designed and developed by senior management personnel who have extensive experience of project management within the oil and gas industry. Not only do they have knowledge of sound management practice and experience of management tools and software programmes in use, the have the added advantage of first-hand knowledge of possible scenarios which affect the smooth running of projects. The ability to recall lessons learned adds a lot more value to the offering than reliance on teaching standard management activities; notably how to manage 'what-if' situations. By interaction and open discussion with the course students, a better understanding of the true life activities of a project can be identified and shared.

The courses developed to date include a basic module and an advanced module, both of which are three days duration and are either delivered at the Clients facility or at a number of convenient global locations. The basic module provides an introduction to the role and required capabilities of the project manager and are targeted for those individuals who an organisation believes will progress to project management and leadership roles. Alternatively, it could be individuals who are already in those roles but are in need of training and refreshment and/or assessment of their suitability.

The advanced module would include those individuals who are already performing well and will be asked to take bigger and more complex challenges. It also provides an opportunity for a work based issue to be researched and examined in detail by the delegate. The findings will then be discussed and further interrogated during attendance of the course.

By adopting this method of teaching, it is anticipated that all students will gain a more focussed and detailed knowledge of project management within the oil & gas and marine sectors. Methods and tools discussed as key elements of successful management will be on specific items linked to these industries. Further development is underway for similar courses in other key industries including building & construction and transport and infrastructure.

5. PASSING THE TESTS

The University agreed that there are currently a range of training packages that are very broad based and are not specific to the energy industry, engineering, construction based capital projects or an organisation's own personal development plans. It was at this point the University considered this newly developed product could be of significant value particularly when coupled with Stapleton International's reputation in the market place. Particularly in respect of their ongoing consultancy support provided to clients within this area.

Following initial development work with the University, a meeting was held with responsible parties to address in detail the introduction of such a course to the University programme. Considering a key element would be that Stapleton International would be responsible for the course delivery, the practicalities of that happening was understandably a concern of the University regulatory authorities at that time. Also there was a necessity to ensure that all courses developed and delivered have a seamless progression and that requisite accreditation and recognition from authorised institutions is fundamental to the success of this initiative. All training materials must also be fully compliant with all relevant national and international governing, regulatory and ISO standards.
One of the factors considered at this time was the potential for students to progress to gaining credits to be used at a later stage in the study of post graduate qualifications. After due consideration this option was deferred in favour of a non-credit bearing programme. It has been agreed that the University will consider how students can progress from the short course onto their degree programmes.

The CV's of Stapleton International staff who would be responsible for the delivery of the programmes were reviewed and approved by University personnel. Following this approval stage, the development of the courses began in earnest by Stapleton International under the watchful eye of senior management and representatives of the University.

Following the development of the course elements by Stapleton International and the collation of training materials, further joint reviews were undertaken with University staff and a summary document was then prepared to ensure compliance with the requirements of QAA. This was a critical stage for the collaboration agreement and guidance was given by the University allowing Stapleton International to ensure that the course content was fully compliant.

5.1 QAA KEY COMPLIANCE ISSUES

5.1 (a) Basic Level Courses

Basic level courses will be at an academic level equivalent to QAA FHEQ Level 4. Students will have demonstrated:

- knowledge of the underlying concepts and principles associated with their area(s) of study, and an ability to evaluate and interpret these within the context of that area of study
- an ability to present, evaluate and interpret qualitative and quantitative data, in order to develop lines of argument and make sound judgements in accordance with basic theories and concepts of their subject(s) of study.

Typically, students will be able to:

- evaluate the appropriateness of different approaches to solving problems related to their area(s) of study and/or work
- communicate the results of their study/work accurately and reliably, and with structured and coherent arguments
- undertake further training and develop new skills within a structured and managed environment.

And will have:

- the qualities and transferable skills necessary for employment requiring the exercise of some personal responsibility.

Students will have a sound knowledge of the basic concepts of a subject, and will have learned how to take different approaches to solving problems. They will be able to communicate accurately and will have the qualities needed for employment requiring the exercise of some personal responsibility. It is expected that these courses may persuade some students to consider obtaining higher level formal qualifications.

5.1 (b) Advanced Level Courses

Advanced level courses will be at an academic level equivalent to QAA FHEQ Level 6. Students will have demonstrated:

- a systematic understanding of key aspects of their field of study, including acquisition of coherent and detailed knowledge, at least some of which is at, or informed by, the forefront of defined aspects of a discipline
- an ability to deploy accurately established techniques of analysis and enquiry within a discipline
- conceptual understanding that enables the student:
    - to devise and sustain arguments, and/or to solve problems, using ideas and techniques, some of which are at the forefront of a discipline
    - to describe and comment upon particular aspects of current research, or equivalent advanced scholarship, in the discipline
- an appreciation of the uncertainty, ambiguity and limits of knowledge
- the ability to manage their own learning, and to make use of scholarly reviews and primary sources (for example, refereed research articles and/or original materials appropriate to the discipline).

Typically, students will be able to:

- apply the methods and techniques that they have learned to review, consolidate, extend and apply their knowledge and understanding, and to initiate and carry out projects
- critically evaluate arguments, assumptions, abstract concepts and data (that may be incomplete), to make judgements, and to frame appropriate questions to achieve a solution - or identify a range of solutions - to a problem
- communicate information, ideas, problems and solutions to both specialist and non-specialist audiences.

And will have:

- the qualities and transferable skills necessary for employment requiring:
- the exercise of initiative and personal responsibility
- decision-making in complex and unpredictable contexts
- the learning ability needed to undertake appropriate further training of a professional or equivalent

5.1 (c) Academic Board Approvals

Following the acceptance by the University that elements of QAA were satisfied, the preparation of the Outline Programme Approval Form and was submitted to the Academic Development Committee. It was recommended that a comprehensive Due Diligence Audit of Stapleton International was undertaken by the University.

Although considered a formality, Stapleton International were surprised at the depth of investigation levels adopted. It was ironic to be at the opposite side to such an activity as it is a core capability of Stapleton International and the methodology was interesting to those involved.

Following on from the Due Diligence, a draft agreement was submitted by the University to Stapleton International for review and comment. Once this was review was completed and the resultant comments were accepted by the University Legal Department, a Programme Agreement was drawn up and signed by respective parties.

6. TRANSFERABILITY

The current courses have been developed as a result of market forces and a shortage of skills within the oil & gas and marine sectors. As stated several times above, this was resultant from the experiences of Stapleton International during the completion of recent assignments primarily completed in the oil & gas and marine sectors. However further research since development started on these courses has revealed that there are other sectors who require a similar facility. There have been a number of key observations by Stapleton International staff which have promoted a further review of input to, and delivery of, short course training programme development.

Our involvement with recent claims in Western Australia have prompted discussions with Clients based around the lack of experienced personnel in other sectors, including construction and infrastructure. There are significant numbers of staff who are well qualified academically and are perhaps lacking skills and experience. An example involved the reporting of schedule slippage on a specific subcontract, the management of which had been delegated to a junior Project Engineer. Reviewing the lack of progress achieved, decisions were made on the presumption that recovery could be accomplished by increased working hours and a change of shift pattern. However, the inability to identify the root cause of slippage was as a result of material quality deficiency simply exacerbated the problem of schedule slippage. This oversight occurred when senior management resources were stretched and consequently the junior engineer was charged with overseeing this element of the project. The progress recovery decision was an error which can be directly attributed to a lack of experience. The fact it was not resolved early demonstrates the requirement for the additional tutoring of practical skills.

It is probable that such occurrences are wide spread and are not being addressed due to the lack of vision of some senior managers. It is for this reason that Stapleton International and the University are reviewing other major sectors as targets for developing similar Project Management short courses as those recently completed for oil & gas and marine. The core principles will remain the same, however the discipline element will be reviewed for sector specific differences.

In the near future, such opportunities will be the subject of a series of workshops between the two parties. Where necessary it will be agreed that input is required from sector experts, experienced senior managers will be called upon to provide detailed input and development of these courses. They will also be required to assist in the delivery of courses following identification of market demand. The resultant courses will establish the University of Sunderland as a leading resource for delivery of corporate and professional short course education.

7. CONCLUSIONS

The opportunities for the development and delivery of short courses has been a challenge for both Stapleton International and the University of Sunderland. Regardless of which company and academic institution set out on a collaboration agreement, the core requirements of each party to retain their performance standards and respective and regulatory compliance are different. It is fundamental therefore that a trust model is achieved at the earliest opportunity.

Knowledge of the individuals involved in the course development and approval is certainly a welcome advantage and the respect of specific knowledge and experience was a prime accelerant in the process. Stapleton International benefited by having resources available who have significant experience of their sector and also many years of experience in teaching and lecturing at degree level. The collaboration was also well served by the existing working relationships and arrangements between both parties.

Commercial arrangements were not considered a core driver of reaching agreement and in practice was probably the easiest issue to resolve. The University have a standard band width; Stapleton International would include this element in their costs in the business plan, hence agreement was swift. This however should not be accepted as being normal with all companies and institutions as many others may have different economic drivers.

The development of a course to ensure alignment with Academic Board requirements, particularly with respect to regulatory compliance is lengthy, if not somewhat complex in some areas. Progress of course content has to be undertaken with strict cognisance to the requirements of QAA. Without this consideration it is impossible to achieve approval, however it did provide the opportunity to better define the structure of the course elements.

Although the initial courses are designed to suit the oil & gas and marine sector skills shortfalls identified, the opportunity to expand into more sectors has already been accepted by both parties. Work is underway on the design and content of future courses within other sectors and both parties are also working diligently in developing the strategy for the supplementary courses. From experience such courses involving elements for effective team working for example, will be required by many of the new projects in prospect, particularly in emerging the African oil &gas markets.

In summary, the progress to reach a collaboration agreement between Stapleton International and the University has been welcomed by both parties and also their Clients. The process is not quick by any measure. It can be difficult and delayed as many elements have to be methodically addressed and agreed; often with differing requirements for each party. However, development of the resultant agreement generated an excellent working relationship. This in turn has provided a solid structure for the increase of aligned business activities, whilst identifying additional opportunities for future collaborative agreements.

8. REFERENCES

1. CULLEN, E., 'Tackling the Skills Shortage Challenge in the Oil, Gas Industry' Rigzone 2015

2. ENZER, G., 'Oil & gas is facing a severe talent shortage' Arabian Oil and Gas 2014

3. SAVITCH, W.J., 'Pascal: An Introduction to the Art and Science of Programming' 1986

4. PICKLES, S., 'Tackling the Skill Shortage' Business News, WA 2014

5. ENZER, G., 'Oil & gas is facing a severe talent shortage' Arabian Oil and Gas 2014

6. http://www.mastersportal.eu/studies/40422/project-management.html

7. Project Management Institute, A Guide to the Project Management, Body of Knowledge 2013

9. AUTHOR BIOGRAPHY

Dr Mick Thurlbeck holds the position of Managing Director of Stapleton International. He is responsible for formulating and successfully implementing company policy, directing strategy towards the profitable growth and operation of the company. Developing strategic operating plans that reflect the longer-term objectives and established priorities.

He is Visiting Fellow in Enterprise at the University of Sunderland within the Faculty of Business and Law and within the Research in Innovation, Sustainability and Enterprise. He is responsible for providing input to research projects and research development, teaching programmes and curriculum development. Also promoting the Academic Areas of the University in Sunderland, the wider region and internationally, particularly Singapore, Malaysia, China and Korea.

His experience includes being a Chartered Fellow of the Chartered Institute of Management and a Fellow of the Royal Institution of Naval Architects

AUTHORS' CONTACT DETAILS

Armando Shinohara
UFPE - Federal University of Pernambuco
shinohara@ufpe.br

Ayako Ono
UFPE - Federal University of Pernambuco
shinoharaah@yahoo.com

Hughes Rodolfo A. Brigoli
University of Cebu
jacklynantolihao@gmail.com

Iman Fiqrie B Muhammad
Malaysian Maritime Academy (ALAM)
wilfredo.yutuc@alam.edu.my

K Sivaprasad
CUSAT
gauthi.shiptech36@gmail.com

Ketut Utama
Institute of Technology Sepuluh Nopember (ITS)
kutama@na.its.ac.id

Mark Symes
Australian Maritime College
m.symes@amc.edu.au

Mick Thurlbeck
Stapleton International Pte Limited
mick.thurlbeck@stapletonpte.com

Richard J. Burke
Maritime College, State University of New York
RBurke@sunymaritime.edu

Richard Thain
Marine Learning Alliance
Richard.Thain@mla-uk.com

The Royal Institution of Naval Architects
8-9 Northumberland Street
London, WC2N 5DA
United Kingdom

ISBN 978-1-5108-8334-5